The Mathemagician and Pied Puzzler

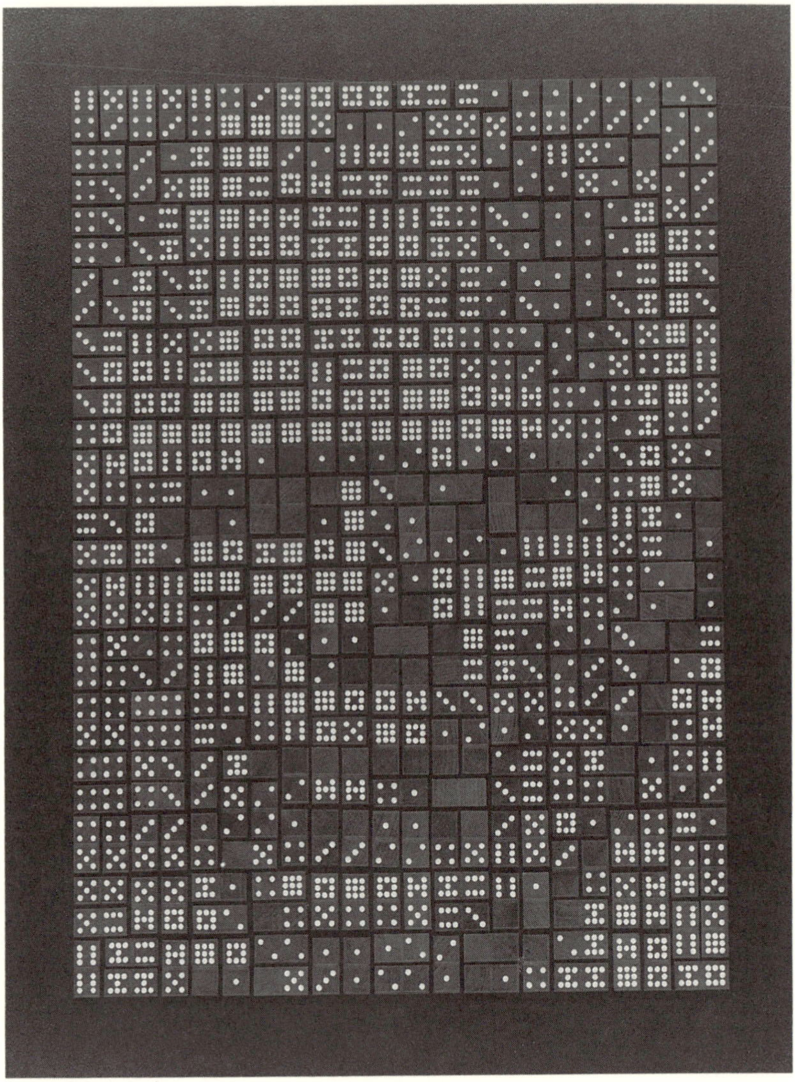

Inspired by Martin Gardner: An arrangement of six complete sets of double nine dominoes. Presented to him as a gift at the 1993 "Gathering for Gardner" in Atlanta, Georgia, by the artist, Ken Knowlton.

The Mathemagician and Pied Puzzler

A Collection in Tribute to Martin Gardner

Edited by Elwyn Berlekamp and Tom Rodgers

A K Peters
Natick, Massachusetts

Editorial, Sales, and Customer Service Office

A K Peters, Ltd.
63 South Avenue
Natick, MA 01760

Library of Congress Cataloging-in-Publication Data

The mathemagician and pied puzzler : a collection in tribute to Martin Gardner / edited by Elwyn Berlekamp and Tom Rodgers.
p. cm.
Includes bibliographical references.
ISBN 1-56881-075-X
1. Mathematical recreations. I. Gardner, Martin. II. Berlekamp, Elwyn. III. Rodgers, Tom.
QA95.M366 1999
793. 7'4–dc21 98-51744
CIP

Printed in the United States of America
03 02 01 00 99 10 9 8 7 6 5 4 3 2 1

Contents

Foreword

Martin Gardner has had no formal education in mathematics, but he has had an enormous influence on the subject. His writings exhibit an extraordinary ability to convey the essence of many mathematically sophisticated topics to a very wide audience. In the words first uttered by mathematician John Conway, Gardner has brought "more mathematics, to more millions, than anyone else."

In January 1957, Martin Gardner began writing a monthly column called "Mathematical Game" in *Scientific American.* He soon became the influential center of a large network of research mathematicians with whom he corresponded frequently. On browsing through Gardner's old columns, one is struck by the large number of now-prominent names that appear therein. Some of these people wrote Gardner to suggest topics for future articles; others wrote to suggest novel twists on his previous articles. Gardner personally answered all of their correspondence.

Gardner's interests extend well beyond the traditional realm of mathematics. His writings have featured mechanical puzzles as well as mathematical ones, Lewis Carroll, and Sherlock Holmes. He has had a life-long interest in magic, including tricks based on mathematics, on sleight of hand, and on ingenious props. He has played an important role in exposing charlatans who have tried to use their skills not for entertainment but to assert supernatural claims. Although he nominally retired as a regular columnist at *Scientific American* in 1982, Gardner's prolific output has continued.

Martin Gardner's influence has been so broad that a large percentage of his fans have only infrequent contacts with each other. Tom Rodgers conceived the idea of hosting a weekend gathering in honor of Gardner to bring some of these people together. The first "Gathering for Gardner" (G4G1) was held in January 1993. Elwyn Berlekamp helped publicize the idea to mathematicians. Mark Setteducati took the lead in reaching the magicians. Tom Rodgers contacted the puzzle community. The site chosen was Atlanta, partly because it is within driving distance of Gardner's home.

The unprecedented gathering of the world's foremost magicians, puzzlists, and mathematicians produced a collection of papers assembled by Scott Kim, distributed to the conference participants, and presented to

Gardner at the meeting. G4G1 was so successful that a second gathering was held in January 1995 and a third in January 1998. As the gatherings have expanded, so many people have expressed interest in the papers presented at prior gatherings that A K Peters, Ltd., has agreed to publish this archival record. Included here are the papers from G4G1 and a few that didn't make it into the initial collection.

The success of these gatherings has depended on the generous donations of time and talents of many people. Tyler Barrett has played a key role in scheduling the talks. We would also like to acknowledge the tireless effort of Carolyn Artin and Will Klump in editing and formatting the final version of the manuscript. All of us felt honored by this opportunity to join together in this tribute to the man in whose name we gathered and to his wife, Charlotte, who has made his extraordinary career possible.

Elwyn Berlekamp
Berkeley, California

Tom Rodgers
Atlanta, Georgia

Part I

Personal Magic

Martin Gardner: A "Documentary"

Dana Richards

> I've never consciously tried to keep myself out of anything I write, and I've always talked clearly when people interview me. I don't think my life is too interesting. It's lived mainly inside my brain. [21]

While there is no biography of Martin Gardner, there are various interviews and articles about Gardner. Instead of a true biography, we present here a portrait in the style of a documentary. That is, we give a collection of quotes and excerpts, without narrative but arranged to tell a story.

The first two times Gardner appeared in print were in 1930, while a sixteen-year-old student at Tulsa Central High. The first, quoted below, was a query to "The Oracle" in Gernsback's magazine *Science and Invention.* The second was the "New Color Divination" in the magic periodical *The Sphinx,* a month later. Also below are two quotes showing a strong childhood interest in puzzles. The early interest in science, magic, puzzles, and writing were to stay with him.

* * *

"I have recently read an article on handwriting and forgeries in which it is stated that ink eradicators do not remove ink, but merely bleach it, and that ink so bleached can be easily brought out by a process of 'fuming' known to all handwriting experts. Can you give me a description of this process, what chemicals are used, and how it is performed?" [1]

* * *

"Enclosed find a dollar bill for a year's subscription to *The Cryptogram.* I am deeply interested in the success of the organization, having been a fan for some time." [2]

* * *

An able cartoonist with an adept mind for science. [1932 yearbook caption.]

* * *

[1934] "As a youngster of grade school age I used to collect everything from butterflies and house keys to match boxes and postage stamps – but when I grew older ... I sold my collections and chucked the whole business, and began to look for something new to collect. Thus it was several years ago I decided to make a collection of mechanical puzzles....

"The first and only puzzle collector I ever met was a fictitious character. He was the chief detective in a series of short stories that ran many years ago in one of the popular mystery magazines.... Personally I can't say that I have reaped from my collection the professional benefit which this man did, but at any rate I have found the hobby equally as fascinating." [3]

* * *

"My mother was a dedicated Methodist who treasured her Bible and, as far as I know, never missed a Sunday service unless she was ill. My father, I learned later, was a pantheist.... Throughout my first year in high school I considered myself an atheist. I can recall my satisfaction in keeping my head upright during assemblies when we were asked to lower our head in prayer. My conversion to fundamentalism was due in part to the influence of a Sunday school teacher who was also a counselor at a summer camp in Minnesota where I spent several summers. It wasn't long until I discovered Dwight L. Moody ... [and] Seventh-Day Adventist Carlyle B. Haynes.... For about a year I actually attended an Adventist church.... Knowing little then about geology, I became convinced that evolution was a satanic myth." [22]

* * *

Gardner was intrigued by geometry in high school and wanted to go to Caltech to become a physicist. At that time, however, Caltech accepted undergraduates only after they had completed two years of college, so Gardner went to the University of Chicago for what he thought would be his first two years.

That institution in the 1930s was under the influence of Robert Maynard Hutchins, who had decreed that everyone should have a broad liberal education with no specialization at first. Gardner, thus prevented from pursuing math and science, took courses in the philosophy of science and then in philosophy, which wound up displacing his interest in physics and Caltech. [19]

* * *

"My fundamentalism lasted, incredibly, through the first three years at the University of Chicago, then as now a citadel of secular humanism.... I was one of the organizers of the Chicago Christian Fellowship.... There was no particular day or even year during which I decided to stop calling myself a Christian. The erosion of my beliefs was even slower than my conversion. A major influence on me at the time was a course on comparative religions taught by Albert Eustace Haydon, a lapsed Baptist who became a well-known humanist." [22]

"After I had graduated and spent another year at graduate work, I decided I didn't want to teach. I wanted to write." [24]

* * *

Gardner returned to his home state after college to work as assistant oil editor for the *Tulsa Tribune.* "Real dull stuff," Gardner said of his reporting stint. He tired of visiting oil companies every day, and took a job ... in Chicago. [17]

* * *

He returned to the Windy City first as a case worker for the Chicago Relief Agency and later as a public-relations writer for the University of Chicago. [9]

* * *

[1940] A slim, middling man with a thin face saturnined by jutting, jetted eyebrows and spading chin, his simian stride and posture is contrasted by the gentility and fluent deftness of his hands. Those hands can at any time be his passport to fame and fortune, for competent magicians consider him one of the finest intimate illusionists in this country today. But to fame Gardner is as indifferent as he is to fortune, and he has spent the last half-dozen years of his life eliminating both from his consideration.

In a civilization of property rights and personal belongings, Martin Gardner is a Robinson Crusoe by choice, divesting himself of all material things to which he might be forced to give some consideration. The son of a well-to-do Tulsa, Oklahoma, family that is the essence of upper middle-class substantiality, Gardner broke from established routine to launch himself upon his self-chosen method of traveling light through life.

Possessor a few years ago of a large, diversified, and somewhat rarefied library, Martin disposed of it all, after having first cut out from the important books the salient passages he felt worth saving or remembering. These clippings he mounted, together with the summarized total of his knowledge, upon a series of thousands of filing cards. Those cards, filling some twenty-five shoe boxes, are now his most precious, and almost only possession. The card entries run from prostitutes to Plautus – which is not too far – and from Plato to police museums.

Chicagoans who are not too stultified to have recently enjoyed a Christmas-time day on Marshall Field and Company's toy floor may remember Gardner as the "Mysto-Magic" set demonstrator for the past two years. He is doing his stint again this season. The rest of the year finds him periodically down to his last five dollars, facing eviction from the Homestead Hotel, and triumphantly turning up, Desperate Desmond fashion, with fifty or a hundred dollars at the eleventh hour – the result of having sold an idea for a magic trick or a sales-promotion angle to any one of a half-dozen companies who look to him for specialties. During the past few months a determined outpouring of ideas for booklets on paper-cutting and other tricks, "pitchmen's" novelties, straight magic and card tricks, and occasional dabblings in writings here and there have made him even more well known as an "idea" man for small novelty houses and children's book publishers....

To Gardner's family his way of life has at last become understandable, but it has taken world chaos to make his father say that his oldest son is perhaps the sanest of his family....

His personal philosophy has been described as a loose Platonism, but he doesn't like being branded, and he thinks Plato, too, might object with sound reason. If he were to rest his thoughts upon one quotation it would be Lord Dunsany's: "Man is a small thing, and the night is large and full of wonder." [5]

* * *

Martin Gardner '36 is a professional [sic] magician. He tours the world pulling rabbits out of hats. When Professor Jay Christ (Business Law) was exhibiting his series of puzzles at the Club late last Fall Gardner chanced to be in town and saw one of the exhibits. He called up Mr. Christ and asked if he might come out to Christ's home. He arrived with a large suitcase full of puzzles! Puzzles had been a hobby with him, but where to park them while he was peregrinating over the globe was a problem. Would Mr. Christ, who had the largest collection he had ever heard of, accept Mr. Gardner's four or five hundred? [4]

* * *

He was appointed yeoman of the destroyer escort in the North Atlantic "when they found out I could type."

"I amused myself on nightwatch by thinking up crazy plots," said the soft-spoken Gardner. Those mental plots evolved into imaginative short stories that he sold to *Esquire* magazine. Those sales marked a turning point in Gardner's career. [18]

* * *

His career as a professional writer started in 1946 shortly after he returned from four years on a destroyer escort in World War II. Still flush with mustering-out pay, Gardner was hanging around his alma mater, the University of Chicago, writing and taking an occasional GI Bill philosophy course. His break came when he sold a humorous short story called "The Horse on the Escalator" to *Esquire* magazine, then based in Chicago. The editor invited the starving writer for lunch at a good restaurant.

"The only coat I had," Gardner recalls, "was an old Navy pea jacket that smelled of diesel oil. I remember the hatcheck girl looking askance when I handed her the filthy rag." [15]

Every Saturday a group of conjureres would gather in a restaurant in lower Manhattan. "There would be 50 magicians or so, all doing magic tricks," Gardner reminisces. One of them intrigued him with a so-called hexaflexagon – a strip of paper folded into a hexagon, which turns inside out when two sides are pinched. Fascinated, Gardner drove to Princeton, where graduate students had invented it. [23]

* * *

He got into mathematics by way of paper folding, which was a big part of the puzzle page at *Humpty*. A friend showed him a novel way to fold a strip of paper into a series of hexagons, which led to an article on combinatorial geometry in *Scientific American* in December 1956. James R. Newman's *The World of Mathematics* had just been published, demonstrating the appeal of math for the masses, and Gardner was asked to do a monthly column. "At the time, I didn't own a single math book," he recalls. "But I knew of some famous math books, and I jumped at the chance." His first columns were simple. Through the years they have grown far more sophisticated in logic, but the mathematics in them has never gone much beyond second-year college level, because that's all the mathematics Gardner knows. [16]

* * *

"*The Annotated Alice*, of course, does tie in with math, because Lewis Carroll was, as you know, a professional mathematician. So it wasn't really too far afield from recreational math, because the two books are filled with all kinds of mathematical jokes. I was lucky there in that I really didn't have anything new to say in *The Annotated Alice* because I just looked over the literature and pulled together everything in the form of footnotes. But it was a lucky idea because that's been the best seller of all my books." [14]

* * *

At first, Gardner says, the column was read mostly by high school students (he could tell by the mail), but, gradually, as he studied the enormous literature on recreational math and learned more about it, he watched his readers become more sophisticated. "This kind of just happened," he explains with a shrug and a gesture toward the long rows of bookshelves, crammed with math journals in every language, that line one alcove in his study. "I'm really a journalist."

Gardner says he never does any original work, he simply popularizes the work of others. "I've never made a discovery myself, unless by accident. If you write glibly, you fool people. When I first met Asimov, I asked him if he was a professor at Boston University. He said no and .... asked me where I got my Ph.D. I said I didn't have one and he looked startled. 'You

About 1947, he moved to New York where he soon became friends such well-known magic devotees as the late Bruce Elliot, Clayton Raw Paul Curry, Dai Vernon, Persi Diaconis, and Bill Simon. It was Simon w introduced Martin and Charlotte (Mrs. Gardner) and served as best mar their wedding. Judge George Starke, another magic friend, performed t ceremony. [12]

* * *

"Ever since I was a boy, I've been fascinated by crazy science and such things as perpetual motion machines and logical paradoxes. I've always enjoyed keeping up with those ideas. I suppose I really didn't get into it seriously until I wrote my first book, *Fads and Fallacies in the Name of Science.* I was influenced by the Dianetics movement, now called Scientology, which was then promoted by John Campbell in *Astounding Science Fiction.* I was astonished at how rapidly the thing had become a cult. I had friends who were sitting in Wilhelm Reich's orgone energy accumulators. And the Immanuel Velikovsky business had just started, too. I wrote about those three things in an article for the *Antioch Review,* then expanded that article into a book by adding chapters on dowsing, flying saucers, the hollow-earth theories, pyramidology, Atlantis, early ESP research, and so on. It took a long time for the book to start selling, but it really took off when they started attacking it on the *Long John Nebel Show....* For about a year, almost every night, the book would be mentioned on the show by some guest who was attacking it." [20]

* * *

Their first son was born in 1955 and their second three years later. Gardner needed a regular income in those years and with his usual serendipity found a job that was just right for him: contributing editor for *Humpty Dumpty's Magazine.* He designed features and wrote stories for *Humpty, Children's Digest, Piggity's,* and *Polly Pigtails.* "Those were good years at *Humpty.*" [15]

* * *

Although Gardner is a brand-new children's writer, he has a good background for the task. He says that he is a great admirer of the L. Frank Baum "Oz" books, having read all of them as a child, and regards Baum as "the greatest writer of children's fiction yet to be produced by America, and one of the greatest writers of children's fantasy in the history of world literature." He adds, "I was brought up on John Martin's magazine, the influence of which can be seen in some of the activity pages which I am contributing to *Humpty Dumpty.*" [6]

* * *

mean you're in the same racket I am,' he said, 'you just read books by the professors and rewrite them?' That's really what I do." [11]

* * *

"I can't think of any definition of 'mathematician' or 'scientist' that would apply to me. I think of myself as a journalist who knows just enough about mathematics to be able to take low-level math and make it clear and interesting to nonmathematicians. Let me say that I think not knowing too much about a subject is an asset for a journalist, not a liability. The great secret of my column is that I know so little about mathematics that I have to work hard to understand the subject myself. Maybe I can explain things more clearly than a professional mathematician can." [20]

* * *

His "Mathematical Games" column in *Scientific American* is one of the few bridges over C. P. Snow's famous "gulf of mutual incomprehesion" that lies between the technical and literary cultures. The late Jacob Bronowski was a devotee; poet W. H. Auden constantly quoted from Gardner. In his novel *Ada,* Vladimir Nabokov pays a twinkling tribute by introducing one Martin Gardiner, whom he calls "an invented philospher."

Nevertheless, as the mathemagician admits, "not all my readers are fans. I have also managed to provoke some outspoken enemies." In the forefront are the credulous victims of Gardner's recent hoaxes: an elaborate treatise that demonstrated the power of pyramid-shaped structures to preserve life and sharpen razor blades, and "proof" by a fictional Dr. Matrix that the millionth digit of π, if it were ever computed, would be the number 5.... Professors at Stanford University have just programmed a computer to carry π to the millionth digit. To everyone's surprise – especially the hoaxer's – the number turned out to be 5. [8]

* * *

"I particularly enjoy writing columns that overlap with philosophical issues. For example, I did a column a few years ago on a marvelous paradox called Newcomb's paradox, in decision theory. It's a very intriguing paradox and I'm not sure that it's even resolved. And then every once in a while I get a sort of scoop. The last scoop that I got was when I heard about a public-key cryptography system at MIT. I realized what a big breakthrough this was and based a column on it, and that was the first publication the general public had on it." [14]

* * *

"I'm very ill at ease in front of an audience," Gardner said. He was asked how he knew he was ill at ease if he had never done it, and that stumped him for a moment. His wife interjected: "The fact is he doesn't want to do it the same way he doesn't want to shop for clothes. To my knowledge he'll shop only for books." [19]

* * *

"My earliest hobby was magic, and I have retained an interest in it ever since. Although I have written no general trade books on conjuring, I have written a number of small books that are sold only in magic shops, and I continue to contribute original tricks to magic periodicals. My second major hobby as a child was chess, but I stopped playing after my college days for the simple reason that had I not done so, I would have had little time for anything else. The sport I most enjoyed watching as a boy was baseball, and most enjoyed playing was tennis. A hobby I acquired late in life is playing the musical saw." [13]

* * *

Gardner himself does not own a computer (or, for that matter, a fax or answering machine). He once did – and got hooked playing chess on it. "Then one day I was doing dishes with my wife, and I looked down and saw the pattern of the chessboard on the surface on the water," he recalls. The retinal retention lasted about a week, during which he gave his computer to one of his two sons. "I'm a scissors-and-rubber-cement man." [23]

* * *

Gardner takes refuge in magic, at which he probably is good enough to earn yet another living. Gardner peers at the world with such wide-eyed wonder as to inspire trust in all who meet him. But when Gardner brings out his green baize gaming board, the wise visitor will keep his money in his pocket. [10]

* * *

"Certain authors have been a big influence on me," Gardner says, and enumerates them. Besides Plato and Kant, there are G. K. Chesterton, William James, Charles S. Peirce, Miguel de Unamuno, Rudolf Carnap, and H. G. Wells. From each Gardner has culled some wisdom. "From Chesterton I got a sense of mystery in the universe, why anything exists," he expounds. "From Wells I took his tremendous interest in and respect for science." ...

"I don't believe God interrupts natural laws or tinkers with the universe," he remarks. From James he derived his notion that belief in God is a matter of faith only. "I don't think there's any way to prove the existence of God logically." [23]

* * *

"In a way I regret spending so much time debunking bad science. A lot of it is a waste of time. I much more enjoyed writing the book with Carnap, or *The Ambidextrous Universe*, and other books about math and science." [26]

* * *

"As a member of a group called the mysterians I believe that we have no idea whether free will exists or how it works.... The mysterians are not an organized group or anything. We don't hold meetings. Mysterians believe that at this point in our evolutionary history there are mysteries that cannot be resolved." [25]

* * *

"There are, and always have been, destructive pseudo-scientific notions linked to race and religion; these are the most widespread and the most damaging. Hopefully, educated people can succeed in shedding light into these areas of prejudice and ignorance, for as Voltaire once said: 'Men will commit atrocities as long as they believe absurdities.' " [7]

* * *

"In the medical field [scientific ignorance] could lead to horrendous results. People who don't understand the difference between a controlled experiment and claims by some quack may die as a result of not taking medical science seriously. One of the most damaging examples of pseudoscience is false memory syndrome. I'm on the board of a foundation exposing this problem." [21]

* * *

"Martin never sold out," Diaconis said. "He would never do anything that he wasn't really interested in, and he starved. He was poor for a very long time until he fit into something. He knew what he wanted to do.... It really is wonderful that he achieved what he achieved." [19]

References

[1] Martin Gardner, "Now It Is Now It Isn't," *Science and Invention*, April 1930, p. 1119.

[2] Martin Gardner, [Letter], *The Cryptogram*, No. 2, April 1932, p. 7.

[3] Martin Gardner, "A Puzzling Collection," *Hobbies*, September 1934, p. 8.

[4] *Tower Topics [University of Chicago]*, 1939, p. 2.

[5] C. Sharpless Hickman, "Escape to Bohemia," *Pulse [University of Chicago]*, vol. 4, no. 1, October 1940, pp. 16–17.

[6] LaVere Anderson, "Under the Reading Lamp," *Tulsa World (Sunday Magazine)*, April 28, 1957, p. 28.

[7] Bernard Sussman, "Exclusive Interview with Martin Gardner," *Southwind [Miami-Dade Junior College]*, vol. 3, no. 1, Fall 1968, pp. 7–11.

[8] [Stefan Kanfer], "The Mathemagician," *Time*, April 21, 1975, p. 63.

[9] Betsy Bliss, "Martin Gardner's Tongue-in-Cheek Science," *Chicago Daily News*, August 22, 1975, pp. 27–29.

[10] Hank Burchard, "The Puckish High Priest of Puzzles," *Washington Post*, March 11, 1976, p. 89.

[11] Sally Helgeson, "Every Day," *Bookletter*, vol. 3, no. 8, December 6, 1976, p. 3.

[12] John Braun, "Martin Gardner," *Linking Ring*, vol. 58, no. 4, April 1978, pp. 47–48.

[13] Anne Commire, "[Martin Gardner]," *Something About the Author*, vol. 16, 1979, pp. 117–119.

[14] Anthony Barcellos, "A Conversation with Martin Gardner," *Two-Year College Mathematics Journal*, vol. 10, 1979, pp. 232–244.

[15] Rudy Rucker, "Martin Gardner, Impresario of Mathematical Games," *Science 81*, vol. 2, no. 6, July/August 1981, pp. 32–37.

[16] Jerry Adler and John Carey, "The Magician of Math," *Newsweek*, November 16, 1981, p. 101.

[17] Sara Lambert, "Martin Gardner: A Writer of Many Interests," *Time-News [Hendersonville, NC]*, December 5, 1981, pp. 1–10.

[18] Lynne Lucas, "The Math-e-magician of Hendersonville," *The Greenville [South Carolina] News*, December 9, 1981, pp. 1B–2B.

[19] Lee Dembart, "Magician of the Wonders of Numbers," *Los Angeles Times*, December 12, 1981, pp. 1, 10–21.

[20] Scot Morris, "Interview: Martin Gardner," *Omni*, vol. 4, no. 4, January 1982, pp. 66–69, 80–86.

[21] Lawrence Toppman, "Mastermind," *The Charlotte [North Carolina] Observer*, June 20, 1993, pp. 1E, 6E.

[22] Martin Gardner, *The Flight of Peter Fromm*, Dover, 1994. Material taken from the Afterword.

[23] Philip Yam, "The Mathematical Gamester," *Scientific American*, December 1995, pp. 38, 40–41.

[24] Istvan Hargittai, "A Great Communicator of Mathematics and Other Games: A Conversation with Martin Gardner," *Mathematical Intelligencer*, vol. 19, no. 4, 1997, pp. 36–40.

[25] Michael Shermer, "The Annotated Gardner," *Skeptic*, vol. 5, no. 2, pp. 56–61.

[26] Kendrick Frazier, "A Mind at Play," *Skeptical Enquirer*, March/April 1998, pp. 34–39.

Ambrose, Gardner, and Doyle

Raymond Smullyan

SCENE I - The year is 2050 A.D.

Professor Ambrose: Have you ever read the book *Science: Good, Bad, and Bogus*, by Martin Gardner?

Professor Byrd: No; I've heard of it, and of course I've heard of Martin Gardner. He was a very famous science writer of the last century. Why do you ask?

Ambrose: Because the book contains one weird chapter – It is totally unlike anything else that Gardner ever wrote.

Byrd: Oh?

Ambrose: The chapter is titled "The Irrelevance of Conan Doyle". He actually advances the thesis that Conan Doyle never wrote the Sherlock Holmes stories–that these stories are forgeries.

Byrd: That *is* weird! Especially from Gardner! On what does he base it?

Ambrose: On absolutely nothing! His whole argument is that no one with the brilliant, rational, scientific mind to write the Sherlock Holmes stories could possibly have spent his last twelve years in a tireless crusade against all rationality–I'm talking about his crazy involvement with spiritualism.

Byrd: To tell you the truth, this fact has often puzzled me! How could anyone with the brilliance to write the Sherlock Holmes stories ever get involved with spiritualism–and in such a crazy way?

Ambrose: You mean that *you* have doubts that Doyle wrote the Holmes stories?

Byrd: Of course not! That thought has never crossed my mind! All I said was that I find the situation *puzzling*. I guess the answer is that Doyle went senile in his last years?

Ambrose: No, no! Gardner correctly pointed out that all the available evidence shows that Doyle remained quite keen and active to the end. He also pointed out that Doyle's interest in spiritualism started much earlier in life than is generally realized. So senility is not the explanation.

Byrd: I just thought of another idea! Perhaps Doyle was planning all along to foist his spiritualism on the public and started out writing his rational Holmes stories to gain everybody's confidence. Then, when the public was convinced of his rationality, whamo!

Ambrose *(After a pause):* That's quite a cute idea! But frankly, it's just as implausible as Gardner's idea that Doyle never wrote the Holmes stories at all.

Byrd: All tight, then; how do *you* explain the mystery?

Ambrose: The explanation is so obvious that I'm amazed that anyone can fail to see it!

Byrd: Well?

Ambrose: Haven't you heard of multiple personalities? Doyle obviously had a dual personality—moreover of a serious psychotic nature! The clue to the whole thing is not *senility* but *psychosis!* Surely you know that some psychotics are absolutely brilliant in certain areas and completely deluded in others. What better explanation could there be?

Byrd: You really believe that Doyle was psychotic?

Ambrose: Of course he was!

Byrd: Just because he believed in spiritualism?

Ambrose: No, his disturbance went much deeper. Don't you know that he believed that the famous Harry Houdini escaped from locked trunks by dematerializing and going out through the keyhole? What's even worse, he absolutely refused to believe Houdini when he said that there was a perfectly naturalistic explanation for the escapes. *Doyle insisted that Houdini was lying!* If that's not psychotic paranoia, what is?

Byrd: I guess you're right. As I said, I never had the slightest doubt that Doyle wrote the Holmes stories, but now your explanation of the apparent contradiction between Doyle the rationalist and Doyle the crank makes some sense.

Ambrose: I'm glad you realize that.

Byrd: But now something else puzzles me: Martin Gardner was no fool; he was surely one of the most interesting writers of the last century. Now, how could someone of Gardner's caliber ever entertain the silly notion that Doyle never wrote the Holmes stories?

Ambrose: To me the solution is obvious: *Martin Gardner never wrote that chapter!* The chapter is a complete forgery. I have no idea *who* wrote it, but it was certainly not Martin Gardner. A person of Gardner's caliber could never have written anything like that!

Byrd: Now just a minute; are you talking about the whole book or just that one chapter?

Ambrose: Just that one chapter. All the other chapters are obviously genuine; they are perfectly consistent in spirit with all the sensible things that Gardner ever wrote. But that one chapter sticks out like a sore thumb—not just with respect to the other chapters, but in relation to all of Gardner's writings. I don't see how there can be the slightest doubt that this chapter is a complete forgery!

Byrd: But that raises serious problems! All right, I can see how an entire book by an alleged author might be a forgery, but an *isolated* chapter of a book? How could the chapter have ever gotten there? Could Gardner have hired someone to write it? That seems ridiculous! Why would he have done a thing like that? On the other hand, why would Gardner have ever allowed the chapter to be included? Or could it possibly have gotten there without his knowledge? That also seems implausible. Will you please explain one thing: *How did the chapter ever get there?* No, your theory strikes me as most improbable!

Ambrose: I agree with you wholeheartedly; the theory *is* most improbable. But the alternative that Gardner actually wrote that chapter is not just improbable, but completely out of the question; he couldn't *possibly* have written such a chapter. And as Holmes wisely said: Whenever we have eliminated the impossible, whatever remains, *however improbable* must be the truth. And so I am forced to the conclusion that Martin Gardner never wrote that chapter. Now, I don't go as far as some historians who believe that Martin Gardner never existed. No, I believe that he did exist, but he certainly never wrote that chapter. We can only hope that future research will answer the question of how that strange chapter ever got into the book. But surely, nobody in his right mind could believe that Gardner actually wrote that chapter.

Byrd *(After a long pause):* I guess you're right. In fact, the more I think about it, you *must* be right! It is certainly not conceivable that anyone as rational as Gardner could entertain such a stange notion. But now I think you've made a very important historical discovery! Why don't you publish it?

Ambrose: I am publishing it. It will appear in the June issue of the *Journal of the History of Science and Literature.* The title is "Gardner and Doyle". I'll send you a copy.

SCENE II - One Hundred Years Later

Professor Broad: Did you get my paper, "Ambrose, Gardner, and Doyle"?

Professor Cranby: No; where did you send it?

Broad: To your Connecticut address.

Cranby: Oh; then I won't get it for a couple of days. What is it about?

Broad: Well, are you familiar with the Ambrose paper, "Gardner and Doyle"?

Cranby: No; I'm familiar with much of Ambrose's excellent work, but not this one. What is it about?

Broad: You know the twentieth century writer, Martin Gardner?

Cranby: Of course! I'm quite a fan of his. I think I have just about everything he ever wrote. Why do you ask?

Broad: Well, you remember his book, *Science: Good, Bad, and Bogus?*

Cranby: Oh, certainly.

Broad: And do you recall the chapter, "The Irrelevance of Conan Doyle"?

Cranby: Oh yes! As a matter of fact that is the strangest chapter of the book and is quite unlike anything else Gardner ever wrote. He seriously maintained that Conan Doyle never wrote the Sherlock Holmes stories.

Broad: Do *you* believe that Doyle wrote the Holmes stories?

Cranby: Of course! Why should I doubt it for one minute?

Broad: Then how do you answer Gardner's objection that no one with a mind so rational as to write the Holmes stories could possibly be so irrational as to get involved with spiritualism in the peculiarly anti-rational way that he did?

Cranby: Oh, come on now! That's no objection! It's obvious that Doyle, with all his brilliance, had an insane streak that simply got worse through the years. Of course, Doyle wrote the Sherlock Holmes stories!

Broad: I heartily agree!

Cranby: The one thing that puzzles me—and I remember that it puzzled me at the time—is how someone like Martin Gardner could ever have believed such an odd-ball thing!

Broad: Ah; that's the whole point of Ambrose's paper! His answer is simply that Gardner never wrote that chapter—the chapter is just a forgery.

Cranby: Good God! That's ridiculous! That's just as crazy as Gardner's idea that Doyle didn't write Holmes. Of course Gardner wrote that chapter!

Broad: Of course he did!

Cranby: But what puzzles me is how such a sober and reliable historian as Ambrose could ever believe that Gardner didn't write that chapter. How could he ever believe anything that bizarre?

Broad: Ah; that's where *my* paper comes in! I maintain that Ambrose never wrote that paper–it must be a complete forgery!

SCENE III - A Hundred Years Later

(To be supplied by the reader)

Discussion: How come this same Martin Gardner, so well known and highly respected for the mathematical games column he wrote for years for *Scientific American,* his numerous puzzle books, his annotated editions of *Alice in Wonderland, The Hunting of the Snark, The Ancient Mariner,* and *Casey at the Bat*–not to mention his religious novel, *The Flight of Peter Fromm,* and his *Whys of a Philosophical Scrivener*–how come he wrote such a crazy chapter as "The Irrelevance of Conan Doyle"?

This troubled me for a long time, until Martin kindly informed me that the whole thing was simply a hoax!

Martin is really great on hoaxes–for example, in his April 1975 column in *Scientific American,* he reported the discovery of a map that required five colors, an opening move in chess (pawn to Queen's rook four) that guaranteed a certain win for white, a discovery of a fatal flaw in the theory of relativity, and a lost manuscript proving that Leonardo da Vinci was the inventor of the flush toilet.

In Martin's book, *Whys and Wherefores* (University of Chicago Press, 1989), is reprinted a scathing review of his *The Whys of a Philosophical Scrivener* by a writer named George Groth. The review ends with the sentence "George Groth, by the way, is one of Gardner's pseudonyms."

A Truth Learned Early

Carl Pomerance

It was in high school that I decided to be a mathematician. The credit (or, perhaps, blame!) for this can be laid squarely on mathematical competitions and Martin Gardner. The competitions led me to believe I had a talent, and for an adolescent unsure of himself and his place in the world, this was no small thing. But Martin Gardner, through his books and columns, led me to the more important lesson that, above all else, mathematics is fun. The contrast with my teachers in school was striking. In fact, there seemed to be two completely different kinds of mathematics: the kind you learned in school and the kind you learned from Martin Gardner. The former was filled with one dreary numerical problem after another, while the latter was filled with flights of fancy and wonderment. From Martin Gardner I learned of logical and language paradoxes, such as the condemned prisoner who wasn't supposed to know the day of his execution (I don't think I understand this even now!), I learned sneaky ways of doing difficult computations (a round bullet shot through the center of a sphere comes to mind), I learned of hexaflexagons (I still have somewhere in my cluttered office a model of a rotating ring I made while in high school), and I learned of islands populated only by truth tellers and liars, both groups being beer lovers. This colorful world stood in stark contrast to school mathematics. I figured that if I could just stick it out long enough, sooner or later I would get to the fun stuff.

It was true; I did get to the fun stuff.

A good part of my job now is being a teacher. Do I duplicate the school experiences I had with my students? Well, I surely try not to, but now I see another side of the story. Technical proficiency is a worthy goal, and when my students need to know, say, the techniques of integration for a later course, I would be remiss if I didn't cover the topic. But I know also that the driving engine behind mathematics is the underlying beauty and power of the subject and that this indeed is the reason it is a subject worth studying. This fundamental truth was learned from Martin Gardner when I was young and impressionable, and it is a truth I carry in my heart. Today, with the national mood for education reform, it seems the rest of the country is finally learning this truth too. Welcome aboard.

Martin Gardner = Mint! Grand! Rare!

Jeremiah Farrell

I was not surprised to discover the wonderful equation in the title (that so aptly describes the person) since logology and numerology are – according to Dr. Matrix – two faces of the same coin. It was Mr. Gardner who introduced me to the wiley doctor over ten years ago. Since then Matrix and I have marveled over the inevitability of Gardner's career choice. His brilliant future as the world's premiere mathematical wordsmith had been fated since the day he was christened. For instance:

(1) There are 13 letters in MARTIN GARDNER. Dr. Matrix notes that 13 is an emirp since its reversal is also prime. The first and last names have six and seven letters. Six is the first perfect number, while seven is the only odd prime that on removal of one letter becomes EVEN. (It can be no coincidence that the even number SIX, upon subtraction of the same letter becomes the odd number IX.)

(2) I had remarked in an issue of *Word Ways* (May, 1981, page 88) that the 3×3 word square in Figure 1 spells out with chess king moves the laudatory phrase "Martin Gardner, an enigma." Not to be outdone (as usual), Matrix has informed me that a better statement is "Martin Gardner: a man and rarer enigma." (He also found that the square had contained a prediction for the 1980 presidential election: "Reagan ran. In!")

I	T	D
G	N	R
M	A	E

Figure 1.

G	A	D
M	E	N
R	I	T

Figure 2.

(3) A different arrangement of the nine letters produces the square shown in Figure 2. Choose three letters, exactly one from each row and one from each column. A common English word will be the result. Dr. Matrix claims this to be an unusual property, not often found in squares composed from names.

This tribute could be continued, but the point is clear. Martin Gardner is indeed Mint! Grand! Rare! He has my love.

Three Limericks — On Space, Time, and Speed

Tim Rowett

Space

Seven steps each ten million to one	10^0 meter = 1 meter = a Human Body
Describe the whole space dimension	10^7 meters = Earth's diameter
The Atom, Cell's girth	10^{14} meters = outer Solar System
Our bodies, the Earth	10^{21} meters = Galaxy's diameter
Sun's System, our Galaxy – done!	10^{28} meters = Universe, and a bit more
	10^{-7} meters = Nucleus of a human cell
	10^{-14} meters = Atom's nucleus

Time

The Creator, seen as an Army Sergeant Major, barks out his orders for the week.

First thing on Monday, Bang!, Light	A week = 7 days corresponds to
Sun and Earth, form up, Friday night	14 Billion Years
At a minute to twelve	1 day corresponds to 2 Billion Years (USA)
Eve spin, Adam delve	1 minute corresponds to 2 Million Years
In the last millisecond, You, right?	1 Millisecond is 23 Years

Speed

Poem	Speed
A child cycles 'round the schoolyard	7 mph – child cyclist
Which lies on the Earth turning hard	700 mph Earth's surface (North Africa)
The Earth rounds the Sun	70,000 mph, rough average speed of Earth
As Sol does "the ton"	700,000 mph turning speed of Galaxy
And our Galaxy flies – Gee! I'm tired	1.4 million mph Galaxy's speed through the debris of the Big Bang

Part II

Puzzlers

A Maze with Rules

Robert Abbott

In his October 1962 column, Martin Gardner presented a puzzle of mine that involved traveling through a city that had various arrows at the intersections. He used another of my puzzles in the November 1963 column – this one involved traveling in three dimensions through a $4 \times 4 \times 4$ grid. At the time I thought these were puzzles, but later I realized they were more like mazes. Around 1980 I started creating more of these things (which I now think could best be described as "mazes with rules"), and in 1990 I had a book of them published, *Mad Mazes.*

The next page shows one of the mazes from my book. This is my manuscript version of the maze, before my publisher added art work and dopey stories. (Actually, I wrote half the dopey stories and I sort of like some of them.) I chose this particular maze because it illustrates the cross-fertilization that Martin's columns created. I got the original idea for this maze from remembering columns that Martin wrote in December 1963, November 1965, and March 1975. These columns presented rolling cube puzzles by Roland Sprague and John Harris. The puzzles involved tipping cubes from one square to another on a grid. As Martin's columns said, you should think of a cube as a large carton that is too heavy to slide but that can be tipped over on an edge.

In my maze, place a die on the square marked START. Position the die so that the 2 is on top and the 6 is facing you (that is, the 6 faces the bottom edge of the page). What you have to do is tip the die off the starting square; then find a way to get it back onto that square. You can tip the die from one square to the next, and you can only tip it onto squares that contain letters. The letters stand for *low, high, odd,* and *even.* If (and only if) a 1, 2, or 3 is on top of the die, then you can tip it onto a square with an *L.* If a 4, 5, or 6 is on top, you can tip it onto a square with an *H.* If a 1, 3, or 5 is on top, you can tip the die onto a square with an *O.* If a 2, 4, or 6 is on top, you can tip the die onto a square with an *E.*

I won't give the solution, but it takes 66 moves.

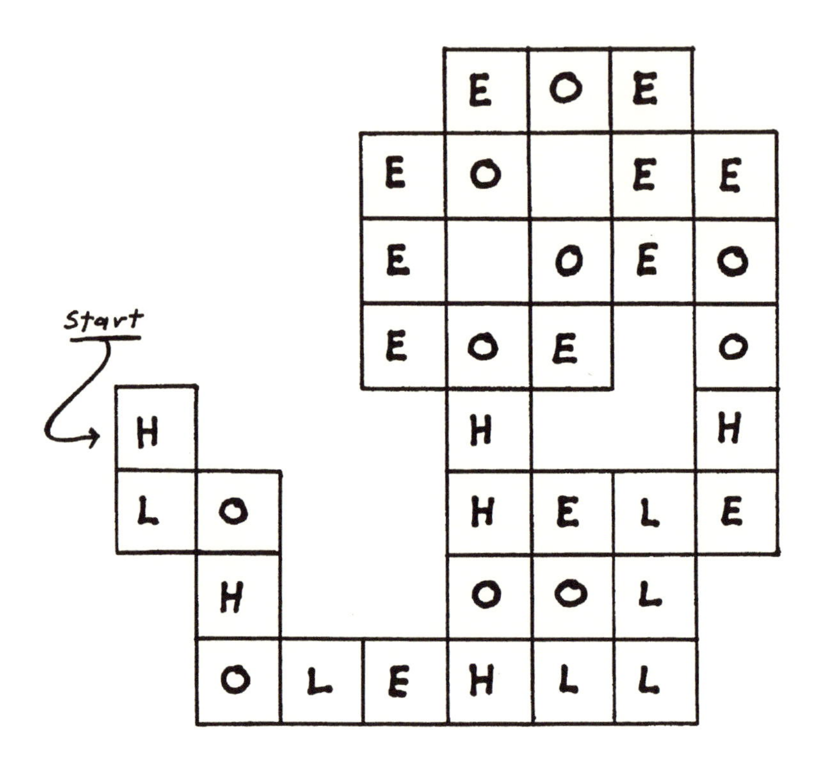

Maze

Addendum, December 1998. Oops! My diagram is too big for this book. The diagram should be at least 6 inches square to have a die roll across it. You might try enlarging it on a copier, but you can also download it off my website. Go to `http://home.att.net/~robtabbott/roll.html`. While you're there, check out the rest of the site. I have a long write-up (with pictures) of something called "walk-through mazes-with-rules." The first of these walk-through mazes appeared at the Gathering for Gardner in January 1993. Since then the concept has grown. In the summer of 1998, several of the mazes were built as adjuncts to large cornfield mazes.

Biblical Ladders

Donald E. Knuth

Charles Lutwidge Dodgson, aka Lewis Carroll, invented a popular pastime now called *word ladders,* in which one word metamorphoses into another by changing a letter at a time. We can go from `THIS` to `THAT` in three such steps: `THIS`, `THIN`, `THAN`, `THAT`.

As an ordained deacon of the Church of England, Dodgson also was quite familiar with the Bible. So let's play a game that combines both activities: Let's construct word ladders in which all words are Biblical. More precisely, the words should all be present in the Bible that was used in Dodgson's day, the King James translation.

Here, for example, is a six-step sequence that we might call Jacob's Ladder, because 'James' is a form of 'Jacob':

... seen of	`JAMES`;	then of all ...	(1 Corinthians 15 : 7)
... because your	`NAMES`	are written ...	(Luke 10 : 20)
... and their	`NAVES`,	and their ...	(1 Kings 7 : 33)
... When the	`WAVES`	of death ...	(2 Samuel 22 : 5)
... had many	`WIVES`.	And his ...	(Judges 8 : 30)
... made their	`LIVES`	bitter with ...	(Exodus 1 : 14)
... And Jacob	`LIVED`	in the land ...	(Genesis 47 : 28)

Puzzle #1. Many people consider the Bible to be a story of transition from `WRATH` to `FAITH`. The following tableau shows, in fact, that there's a Biblical word ladder corresponding to such a transition. But the tableau lists only the verse numbers, not the words; what are the missing words?

(Remember to use a King James Bible for reference, not a newfangled translation! Incidentally, the sequence of verses in this ladder is strictly increasing through the Old Testament, never backtracking in Biblical order; Jacob's Ladder, on the other hand, was strictly decreasing.)

... that his	WRATH	was kindled. ...	(Genesis 39:19)
...	_____	...	(Genesis 40:2)
...	_____	...	(Exodus 24:4)
...	_____	...	(Exodus 34:1)
...	_____	...	(Leviticus 13:3)
...	_____	...	(Leviticus 14:46)
...	_____	...	(Leviticus 25:29)
...	_____	...	(Deuteronomy 22:21)
...	_____	...	(Joshua 11:4)
...	_____	...	(1 Samuel 13:20)
...	_____	...	(1 Samuel 15:3)
...	_____	...	(1 Samuel 26:13)
...	_____	...	(1 Kings 10:15)
...	_____	...	(Psalm 10:14)
...	_____	...	(Isaiah 3:17)
...	_____	...	(Isaiah 54:16)
...	_____	...	(Isaiah 54:17)
... by his	FAITH.	Yea also, ...	(Habakkuk 2:4)

Puzzle #2. Of course #1 was too easy. So neither words nor verse numbers will be given this time. Go from `SWORD` to (plow) `SHARE` in four steps:

... not lift up a	SWORD	against ...	(Micah 4:3)
...	_____	...	( )
...	_____	...	( )
...	_____	...	( )
... every man his	SHARE,	and his ...	(1 Samuel 13:20)

(Hint: In the time of King James, people never swore; they sware.)

Puzzle #3. Of course #2 was also pretty easy, if you have a good concordance or a King James computer file. How about going from `NAKED` to `COVER`, in eight steps? A suitable middle verse is provided as a clue.

... they were	NAKED;	and they ...	(Genesis 3:7)
...	_____	...	( )
...	_____	...	( )
...	_____	...	( )
...	_____	...	(Luke 17:27)
...	_____	...	( )
...	_____	...	( )
...	_____	...	( )
... charity shall	COVER	the multitude ...	(1 Peter 4:8)

Puzzle #4. Find a Biblical word ladder from HOLY to WRIT.

Puzzle #5. (For worshippers of automobiles.) Construct a 12-step Biblical ladder from FORDS (Judges 3:28) to ROLLS (Ezra 6:1).

Puzzle #6. Of course #5 was too hard, unless you have special resources. Here's one that *anybody* can do, with only a Bible in hand. Complete the following Biblical ladder, which "comes back on itself" in an unexpected way.

... seventy times	SEVEN.	Therefore ...	(Matthew 18:22)
...	_____	...	(Matthew 13:)
...	_____	...	(John 4:)
...	_____	...	(Numbers 33:)
...	_____	...	(Deuteronomy 29:)
...	_____	...	(1 Samuel 15:)
...	_____	...	(Job 16:)
...	_____	...	(1 Kings 7:)
...	_____	...	(Romans 9:)
...	_____	...	(Leviticus 9:)
...	_____	...	(Acts 27:)
...	_____	...	(Jeremiah 10:)
...	_____	...	(Matthew 13:)
...	_____	...	(Ezekiel 24:)
...	_____	...	(Matthew 18:)

Puzzle #7. Finally, a change of pace: Construct 5 × 5 word squares, using only words from the King James Bible verses shown. (The words will read the same down as they do across.)

_____	Matthew 11:11	_____	Exodus 28:33
_____	Judges 9:9	_____	Lamentations 2:13
_____	Mark 12:42	_____	1 Peter 5:2
_____	Ecclesiastes 9:3	_____	Genesis 34:21
_____	Luke 9:58	_____	Acts 20:9

Answers

Puzzle #1.

…	`WRATH`	was kindled. …	(Genesis 39:19)
… Pharaoh was	`WROTH`	against two …	(Genesis 40:2)
… And Moses	`WROTE`	all the …	(Exodus 24:4)
… and I will	`WRITE`	upon *these* …	(Exodus 34:1)
… is turned	`WHITE`,	and the …	(Leviticus 13:3)
… all the	`WHILE`	that it is …	(Leviticus 14:46)
… within a	`WHOLE`	year after …	(Leviticus 25:29)
… play the	`WHORE`	in her …	(Deuteronomy 22:21)
… the sea	`SHORE`	in multitude, …	(Joshua 11:4)
… man his	`SHARE`,	and his …	(1 Samuel 13:20)
… have, and	`SPARE`	them not; …	(1 Samuel 15:3)
… a great	`SPACE`	*being* between …	(1 Samuel 26:13)
… of the	`SPICE`	merchants, and …	(1 Kings 10:15)
… mischief and	`SPITE`,	to requite …	(Psalm 10:14)
… Lord will	`SMITE`	with a scab …	(Isaiah 3:17)
… created the	`SMITH`	that bloweth …	(Isaiah 54:16)
… is of me,	`SAITH`	the LORD. …	(Isaiah 54:17)
… by his	`FAITH`.	Yea also, …	(Habakkuk 2:4)

Puzzle #2. Here's a strictly decreasing solution:

… not lift up a	`SWORD`	against …	(Micah 4:3)
… The Lord GOD hath	`SWORN`	by himself, …	(Amos 6:8)
… *sheep that are even*	`SHORN`,	which came …	(Song of Solomon 4:2)
… beside Eloth, on the	`SHORE`	of the Red …	(1 Kings 9:26)
… every man his	`SHARE`,	and his …	(1 Samuel 13:20)

There are 13 other possible citations for `SWORN`, and 1 Kings 4:29 could also be used for `SHORE`, still avoiding forward steps.

Puzzle #3. First observe that the word in Luke 17:27 must have at least one letter in common with both `NAKED` and `COVER`. So it must be `WIVES` or `GIVEN`; and `GIVEN` doesn't work, since neither `GAKED` nor `NIKED` nor `NAVED` nor `NAVEN` is a word. Thus the middle word must be `WIVES`, and the step after `NAKED` must be `WAKED`. Other words can now be filled in.

... they *were*	NAKED;	and they ...	(Genesis 3:7)
... again, and	WAKED	me, as a ...	(Zechariah 4:1)
... which is	WAVED,	and which ...	(Exodus 29:27)
... the mighty	WAVES	of the sea. ...	(Psalm 93:4)
... they married	WIVES,	they were ...	(Luke 17:27)
... hazarded their	LIVES	for the name ...	(Acts 15:26)
... looked in the	LIVER.	At his ...	(Ezekiel 21:21)
... hospitality, a	LOVER	of good men, ...	(Titus 1:8)
... charity shall	COVER	the multitude ...	(1 Peter 4:8)

(Many other solutions are possible, but none are strictly increasing or decreasing.)

Puzzle #4. Suitable intermediate words can be found, for example, in Revelation 3:11; Ruth 3:16; Ezra 7:24; Matthew 6:28; Job 40:17; Micah 1:8; Psalm 145:15. (But 'writ' is not a Biblical word.)

Puzzle #5. For example, use intermediate words found in Matthew 24:35; Ezra 34:25; Acts 2:45; Ecclesiastes 12:11; Matthew 25:33; Daniel 3:21; John 18:18; Genesis 32:15; 2 Corinthians 11:19; Psalm 84:6; Numbers 1:2. (See also Genesis 27:44.)

Puzzle #6.

... times	SEVEN.	Therefore ...	(Matthew 18:22)
... forth, and	SEVER	the wicked ...	(Matthew 13:49)
... seventh hour the	FEVER	left him. ...	(John 4:52)
... and to the	FEWER	ye shall give ...	(Numbers 33:54)
... from the	HEWER	of thy wood ...	(Deuteronomy 29:11)
... And Samuel	HEWED	Agag in pieces ...	(1 Samuel 15:33)
... I have	SEWED	sackcloth ...	(Job 16:15)
... hewed stones,	SAWED	with saws, ...	(1 Kings 7:9)
... remnant shall be	SAVED:	For he will ...	(Romans 9:27)
... shoulder Aaron	WAVED	*for* a wave ...	(Leviticus 9:21)
... violence of the	WAVES.	And the soldiers' ...	(Acts 27:41)
... Gather up thy	WARES	out of ...	(Jeremiah 10:17)
... and sowed	TARES	among the ...	(Matthew 13:25)
... And your	TIRES	*shall be* ...	(Ezekiel 24:23)
... till seven	TIMES?	Jesus saith ...	(Matthew 18:21)

Puzzle #7.

```
WOMEN   BELLS
OLIVE   EQUAL
MITES   LUCRE
EVENT   LARGE
NESTS   SLEEP
```

References

Martin Gardner, *The Universe in a Handkerchief: Lewis Carroll's Mathematical Recreations, Games, Puzzles, and Word Plays* (New York: Copernicus, 1996), Chapter 6.

`http://etext.virginia.edu/kjv.browse.html` [online text of the King James Bible provided in searchable form by the Electronic Text Center of the University of Virginia].

Card Game Trivia

Stewart Lamle

14th Century: Decks of one-sided Tarot playing cards first appeared in Europe. They were soon banned by the Church. (Cards, like other forms of entertainment and gambling, competed with Holy services.) Card-playing spread like wildfire.

16th Century: The four suits were created to represent the ideal French national, unified (feudal) society as promoted by Joan of Arc: Nobility, Aristocracy, Peasants, the Church (Spades, Diamonds, Clubs, Hearts).

18th Century: Symmetric backs and fronts were designed to prevent cheating by signaling to other players.

19th Century: The Joker was devised by a Mississippi riverboat gambler to increase the odds of getting good Poker hands.

20th Century: After 600 years of playing with one-sided cards, two-sided playing cards and games were invented by Stewart Lamle. "Finally, you can play with a full deck!"–Zeus™, Maxx™, and Betto™, are all two-sided card games.

Over 100 million decks of cards were sold in the United States last year!

Problem 11: *N. Yoshigahara*

Using as few cuts as possible, divide the left-hand shape and rearrange the pieces to make the right-hand shape. How many pieces do you need?

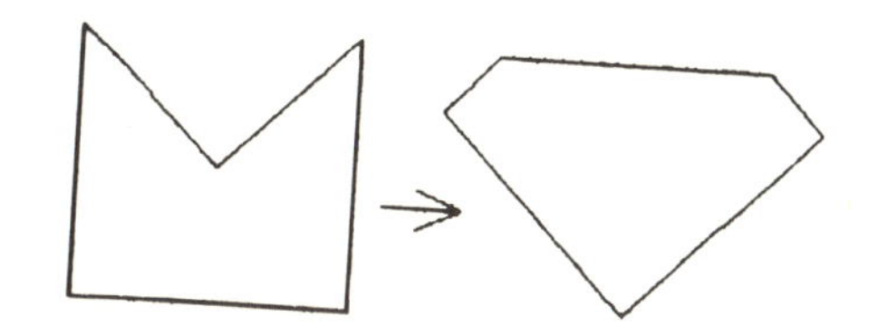

Creative Puzzle Thinking

Nob Yoshigahara

Problem 1:

"An odd number plus an odd number is an even number, and an even number plus an odd number is an odd number. OK?"
"OK."
"An even number plus an even number is an even number. OK?"
"Of course."
"An odd number times an odd number is an odd number, and an odd number times an even number is an even number. OK?"
"Yes."
"Then an even number times an even number is an odd number. OK?"
"No! It is an even number."
"No! It is an odd number! I can prove it!"
How?

Problem 2:

Move two matches so that no triangle remains.

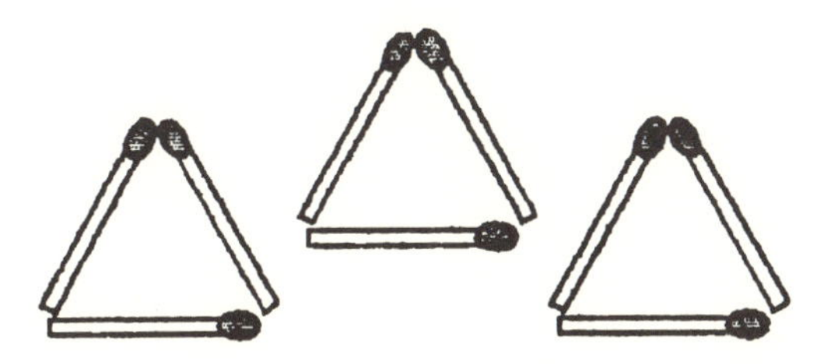

Problem 3:

What number belongs at ? in this sequence?

	24	28	33	34	32	36	35	46	52	53
13	11	17	16	18	14	22	?	33	19	24

Problem 4:

Arrange the following five pieces to make the shape of a star.

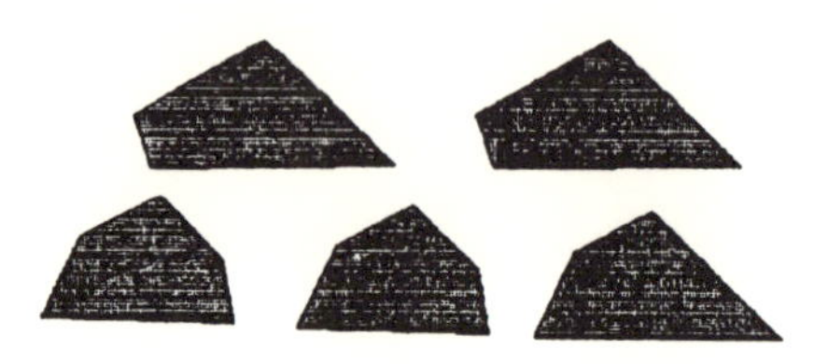

Problem 5:

Calculate the expansion of the following 26 terms.

$$(x-a)(x-b)(x-c)\ldots(x-y)(x-z)$$

Problem 6:

Which two numbers come at the end of this sequence?
2, 4, 6, 30, 32, 34, 36, 40, 42, 44, 46, 50, 52, 54, 56, 60, 62, 64, 66, *x*, *y*

Problem 7:

The figure shown here is the solution to the problem of dividing the figure into four identical shapes. Can you divide the figure into *three* identical shapes?

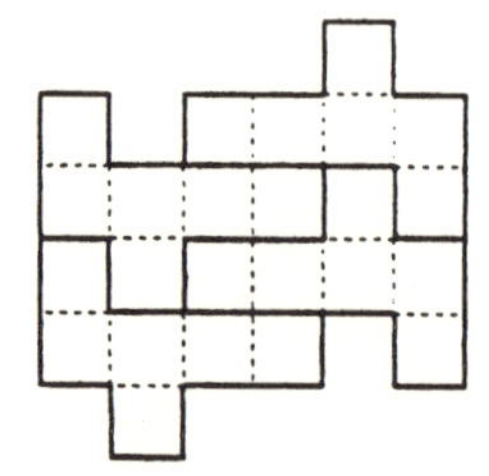

Problem 8:

A 24-hour digital watch has many times that are palindromic. For example, 1:01:01, 2:41:42, 23:55:32, 3:59:53, 13:22:31, etc. (Ignore the colons.) These curious combinations occur 660 times a day.
(1) Find the closest such times.
(2) Find the two palindromes whose difference is closest to 12 hours.
(3) Find the longest time span without a palindromic time.

Problem 9:

A right triangle with sides 3, 4, and 5 matchsticks long is divi
parts with equal area using 3 matchsticks.

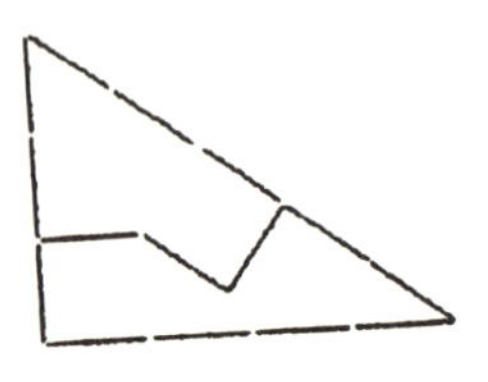

Can you divide this triangle into two parts with equal area using
matchsticks?

Problem 10:

Cover the 7 × 7 square on the left with the 12 L-shaped pieces on t
right. You are not allowed to turn over any of the pieces, but you m
rotate them in the plane.

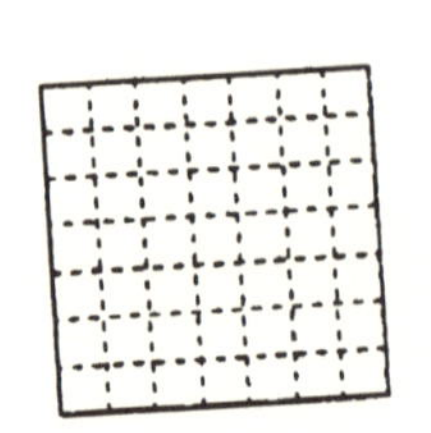

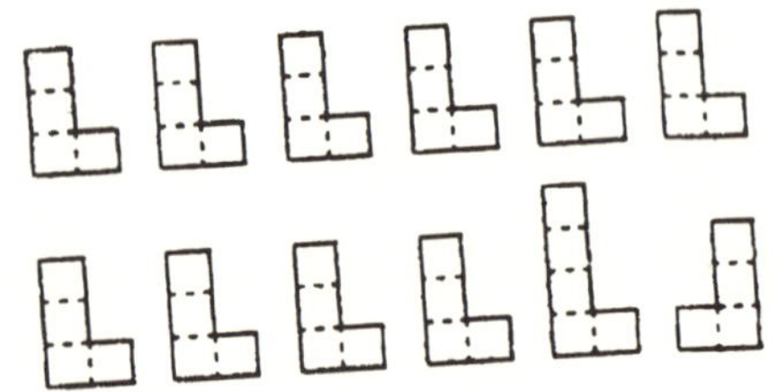

Number Play, Calculators, and Card Tricks: Mathemagical Black Holes

Michael W. Ecker

The legend of Sisyphus is a lesson in inevitability. No matter how Sisyphus tried, the small boulder he rolled up the hill would always come down at the last minute, pulled inexorably by gravity.

Like the legend, the physical universe has strange entities called black holes that pull everything toward them, never to escape. But did you know that we have comparable bodies in recreational mathematics?

At first glance, these bodies may be even more difficult to identify in the world of number play than their more famous brethren in physics. What, after all, could numbers such as 123, 153, 6174, 4, and 15 have in common with each other, as well as with various card tricks?

These are mathematical delights interesting in their own right, but much more so collectively because of the common theme linking them all. I call such individual instances *mathemagical black holes.*

The Sisyphus String: 123

Suppose we start with any natural number, regarded as a string, such as 9,288,759. Count the number of even digits, the number of odd digits, and the total number of digits. These are 3 (three evens), 4 (four odds), and 7 (seven is the total number of digits), respectively. So, use these digits to form the next string or number, 347.

Now repeat with 347, counting evens, odds, total number, to get 1, 2, 3, so write down 123. If we repeat with 123, we get 123 again. The number 123 with respect to this process and the universe of numbers is a mathemagical black hole. All numbers in this universe are drawn to 123 by this process, never to escape.

Based on articles appearing in *(REC) Recreational and Educational Computing.*

But will every number really be sent to 123? Try a really big number now, say 122333444455555666666777777788888888999999999 (or pick one of your own).

The numbers of evens, odds, and total are 20, 25, and 45, respectively. So, our next iterate is 202,545, the number obtained from 20, 25, 45. Iterating for 202,545 we find 4, 2, and 6 for evens, odds, total, so we have 426 now. One more iteration using 426 produces 303, and a final iteration from 303 produces 123.

At this point, any further iteration is futile in trying to get away from the black hole of 123, since 123 yields 123 again. If you wish, you can test a lot more numbers more quickly with a computer program in BASIC or other high-level programming language. Here's a fairly generic one (Microsoft BASIC):

```
  1  CLS
  2  PRINT "The 123 Mathemagical Black Hole / (c) 1993, Dr. M. W. Ecker"
  3  PRINT: PRINT "I'll ask you to input a positive whole number now."
  4  PRINT "I'll count the numbers of even digits, odd digits, and total."
  5  PRINT "From that I'll form the next number.  Surprisingly, we always"
  6  PRINT "wind up reaching the mathemagical black hole of 123...": PRINT
  7  FOR DL=1 TO 1000:NEXT
10  INPUT "What is your initial whole number"; N$: PRINT
20  IF VAL(N$) < 1 OR VAL(N$) < > INT(VAL(N$)) THEN 10
30  FOR DIGIT = 1 TO LEN(N$)
40  D$ = MID$(N$,DIGIT,1)
50  IF D$ = " " THEN 70
60  IF VAL(D$)/2 = INT(VAL(D$)/2) THEN EVEN = EVEN + 1 ELSE ODD = ODD + 1
70  NEXT DIGIT
80  PRINT "EVEN, ODD, TOTAL"
90  NU$ = STR$(EVEN) + STR$(ODD) + STR$(EVEN + ODD)
100  PRINT EVEN;" ";ODD;" ";EVEN + ODD;"---> New number is"; VAL(NU$)
110  PRINT:IF VAL(NU$) = VAL(N$) THEN PRINT "Done.": END
120  N$ = NU$: EVEN = 0: ODD = 0: GOTO 30
```

If you wish, modify line 110 to allow the program to start again. Or revise the program to automate the testing for all natural numbers in some interval.

What Is a Mathemagical Black Hole?

There are two key features that make our example interesting:

1. Once you hit 123, you never get out, just as reaching a black hole of physics implies no escape.

2. Every element subject to the force of the black hole (the process applied to the chosen universe) is eventually pulled into it. In this case, sufficient iteration of the process applied to any starting number must eventually result in reaching 123.

Of course, once drawn in per point 2, an element never escapes, as point 1 ensures.

A *mathemagical black hole* is, loosely, any number to which other elements (usually numbers) are drawn by some stated process. Though the number itself is the star of the show, the real trick is in finding interesting processes.

Formalized Definition. In mathematical terms, a black hole is a triple (b, U, f), where b is an element of a set U and $f: U \rightarrow U$ is a function, all satisfying:

1. $f(b) = b$.
2. For each x in U, there exists a natural number k satisfying $f^k(x) = b$.

Here, b plays the role of the black-hole element, and the superscript indicates k-fold (repeated) composition of functions.

For the Sisyphus String[1], $b = 123$, $U = \{\text{natural numbers}\}$, and $f(\text{number})$ = the number obtained by writing down the string counting # even digits of number, # odd digits, total # digits.

Why does this example work, and why do most mathemagical black holes occur? My argument is to show that large inputs have smaller outputs, thus reducing an infinite universe to a manageable finite one. At that point, an argument by cases, or a computer check of the finitely many cases, suffices.

In the case of the 123 hole, we can argue as follows: If $n > 999$, then $f(n) < n$. In other words, the new number that counts the digits is smaller than the original number. (It's intuitively obvious, but try induction if you would like rigor.) Thus, starting at 1000 or above eventually pulls one down to under 1000. For $n < 1000$, I've personally checked the iterates of $f(n)$ for $n = 1$ to 999 by a computer program such as the one above. The direct proof is actually faster and easier, as a three-digit string for a number must have one of these possibilities for (# even digits, # odd digits, total # digits):

$$(0, 3, 3)$$
$$(1, 2, 3)$$

[1]For generalized sisyphian strings, see *REC*, No. 48, Fall 1992.

(2, 1, 3)
(3, 0, 3)

So, if $n < 1000$, within one iteration you must get one of these four triples. Now apply the rule to each of these four and you'll see that you always produce (1, 2, 3) — thus resulting in the claimed number of 123.

Words to Numbers: 4

Here is one that master recreationist Martin Gardner wrote to tell me about several years ago. Take any whole number and write out its numeral in English, such as FIVE for the usual 5. Count the number of characters in the spelling. In this case, it is 4 — or FOUR. So, work now with the 4 or FOUR. Repeat with 4 to get 4 again.

As another instance, try 163. To avoid ambiguity, I'll arbitrarily say that we will include spaces and hyphens in our count. Then, 163 appears as ONE HUNDRED SIXTY-THREE for a total count of 23. In turn, this gives 12, then 6, then 3, then 5, and finally 4.

Though this result is clearly language-dependent, other natural languages may have a comparable property, but not necessarily with 4 as the black hole.

Narcissistic Numbers: 153

It is well known that, other than the trivial examples of 0 and 1, the only natural numbers that equal the sum of the cubes of their digits are 153, 370, 371, and 407. Of these, just one has a black-hole property.

To create a black hole, we need to define a universe (set U) and a process (function f). We start with any positive whole number that is a multiple of 3. Recall that there is a special shortcut to test whether you have a multiple of 3. Just add up the digits and see whether that sum is a multiple of 3. For instance, 111,111 (six ones) is a multiple of 3 because the sum of the digits, 6, is. However, 1,111,111 (seven ones) is not.

Since we are going to be doing some arithmetic, you may wish to take out a hand calculator and/or some paper. Write down your multiple of 3. One at a time, take the *cube* of each digit. Add up the cubes to form a new number. Now repeat the process. You must reach 153. And once you reach 153, one more iteration just gets you 153 again.

Let's test just one initial instance. Using the sum of the cubes of the digits, if we start with 432 – a multiple of 3 – we get 99, which leads to 1458, then 702, which yields 351, finally leading to 153, at which point future iterations keep producing 153. Note also that this operation or process preserves divisibility by 3 in the successive numbers.

```
10  CLS
20  PRINT "The mathemagical black hole 153...": PRINT
30  PRINT "Simple test program for sum of cubes of digits...": PRINT
40  PRINT "Copyright 1993, Dr. M. W. Ecker"
50  PRINT: FOR DELAY = 1 TO 2000: NEXT
60  INPUT "Number to be tested";N: PRINT
70  IF N/3 > INT(N/3) OR N < 1 THEN PRINT "Positive multiples of 3 only!":
     GOTO 70
80  IF N > 10000000# THEN PRINT "Let's stick to numbers under 10000000":
     GOTO 80
90  N$ = STR$(N): L = LEN(N$): 'Convert to a string to manipulate digits
100  SUMCUBES = 0: 'Initialize sum
110  FOR DIGIT = 1 TO L
120  V(DIGIT) = VAL(MID$(N$,DIGIT,1)): 'Get value of each digit
130  CU(DIGIT) = V(DIGIT)*V(DIGIT)*V(DIGIT): 'Cube each digit
140  SUMCUBES = SUMCUBES + CU(DIGIT): 'Keep running total of sum of cubes
150  NEXT DIGIT
160  PRINT "The sum of the cubes of the digits of " N " is " SUMCUBES
170  IF SUMCUBES = N THEN PRINT "Success! Found a black hole!": PRINT:
     GOTO 70
180  N = SUMCUBES: GOTO 110
```

This program continues forever, so break out after you've grown weary. One nice thing is that it is easy to edit this program to test for black holes using larger powers. (It is well known that none exists for the sum of the *squares* of the digits, as one gets cycles.)

In more formal language, we obtain the 153 mathemagical black hole by letting $U = 3Z^+ = \{$all positive integral multiples of 3$\}$ and $f(n) =$ the sum of the cubes of the digits of n. Then $b = 153$ is the unique black-hole element. (For a given universe, if a black hole exists, it is necessarily unique.)

Not incidentally, this particular result, without the "black hole" terminology or perspective, gets discovered and re-discovered annually, with a paper or problem proposal in one of the smaller math journals every few years.

The argument for why it works is similar to the case with the 123 example. First of all, $1^3 + 3^3 + 5^3 = 153$, so 153 is indeed a fixed point. Second, for the black-hole attraction, note that, for large numbers n, $f(n) < n$. Then, for suitably small numbers, by cases or computer check, each value eventually is "pulled" into the black hole of 153. I'll omit the proof.

To find an analagous black hole for larger powers (yes, there are some), you will need first to discover a number that equals the sum of the fourth (or higher) power of its digits, and then test to see whether other numbers are drawn to it.

Card Tricks, Even

Here's an example that sounds a bit different, yet meets the two criteria for a black hole. It's a classic card trick.

Remove 21 cards from an ordinary deck. Arrange them in seven horizontal rows and three vertical columns. Ask somebody to think of one of the cards without telling you which card he (or she) is thinking of.

Now ask him (or her) which of the three columns contains the card. Regroup the cards by picking up the cards by whole columns intact, *but be sure to sandwich the column that contains the chosen card between the other two columns.* Now re-lay out the cards by laying out by rows (i.e., laying out three across at a time). Repeat asking which column, regrouping cards with the designated column being in the middle, and re-dealing out by rows. Repeat one last time.

At the end, the card chosen must be in the center of the array, which is to say, card 11. This is the card in the fourth row and second column.

There are two ways to prove this, but the easier way is to draw a diagram that illustrates where a chosen card will end up next time. But for those who enjoy programs, try this one from one of my readers.

```
10  CLS: PRINT "The 21 Cards - Program by Sally Frazza
20  PRINT "adapted by Dr. M. W. Ecker for REC
100  DEFINT A-Z
110  DIM V(21), X(7,3)
120  ITER = 0
130  FOR N = 1 TO 21: V(N) = N: NEXT N
140  PRINT: PRINT "Pick a card, please...": PRINT
150  IF ITER = 3 THEN PRINT: PRINT "Your card is"; V(11): END
160  ITER = ITER + 1
170  N = 0
180  FOR I = 1 TO 7: FOR J = 1 TO 3
```

```
190  N = N + 1
200  X(I,J) = V(N)
210  NEXT J: NEXT I
220  PRINT: PRINT
221  FOR I = 1 TO 7
222  PRINT USING "##"; X(I,1); X(I,2); X(I,3)
223  NEXT I
230  PRINT
240  INPUT "Column of card (1, 2, or 3)"; C
250  IF C < 1 OR C > 3 THEN 240
260  FOR K = 1 TO 3: O(K) = K: NEXT K
270  IF C < > 2 THEN SWAP O(C), O(2)
290  N = 0
300  FOR K = 1 TO 3: FOR I = 1 TO 7
310  J = O(K)
320  N = N + 1
330  V(N) = X(I,J)
340  NEXT I: NEXT K
350  GOTO 150
```

Perhaps it is not surprising, but this trick, as with the sisyphian strings, generalizes somewhat.[2]

Kaprekar's Constant: What a Difference 6174 Makes!

Most black holes, nonetheless, involve numbers. Take any four-digit number except an integral multiple of 1111 (i.e., don't take one of the nine numbers with four identical digits). Rearrange the digits of your number to form the largest and smallest strings possible. That is, write down the largest permutation of the number, the smallest permutation (*allowing initial zeros as digits*), and subtract. Apply this same process to the difference just obtained. Within the total of seven steps, you always reach 6174. At that point, further iteration with 6174 is pointless: 7641−1467 = 6174.

Example: Start with 8028. The largest permutation is 8820, the smallest is 0288, and the difference is 8532. Repeat with 8532 to calculate 8532−2358 = 6174. Your own example may take more steps, but you will reach 6174.

---

[2] *REC,* No. 48, Fall 1992.

The Divisive Number 15

Take any natural number larger than 1 and write down its divisors, including 1 and the number itself. Now take the sum of the digits of these divisors. Iterate until a number repeats.

The black-hole number this time is 15. Its divisors are 1, 3, 5, and 15, and these digits sum to 15. This one is a bit more tedious, but it is also that much more strange at the same time. This one may defy not only your ability to explain it, but your very equilibrium.

Fibonacci Numbers: Classic Results as Black Holes

Many an endearing problem has charmed mathophiles with the Fibonacci numbers: 1, 1, 2, 3, 5, 8, 13, 21, 34, 55, 89, ... The first two numbers are each 1 and successive terms are obtained by adding the immediately preceding two elements. More formally, $F(1) = 1 = F(2)$, and $F(n) = F(n-1) + F(n-2)$ for integers $n > 2$.

One of the classic results is that the ratio of consecutive terms has a limiting value. That is, form the ratios $F(n+1)/F(n)$: $\frac{1}{1} = 1\frac{2}{1} = 2\frac{3}{2} = 1.5, \frac{5}{3} = 1.667$ (approx.), $\frac{8}{5} = 1.6$, etc. The ratios seem to be converging to a number around 1.6 or so. In fact, it is well known that the sequence converges to $(1+\sqrt{5})/2$ or the golden number, phi. Its value is roughly 1.618 ... .

Had we used the ratios $F(n)/F(n+1)$ instead, we would have obtained the reciprocal of phi as the limit, but fewer authors use that approach.

To get a quick and dirty feel for why phi arises, but without using a program, *assume* that the numbers $F(n+1)/F(n)$ approach *some* limit — call it L — as n increases. Divide both sides of the equation $F(n) = F(n-1) + F(n-2)$ by $F(n-1)$. For large n this equation is approximately the same as $L = 1 + 1/L$. If we multiply both sides by L we obtain a quadratic equation:

$$(*) \qquad L^2 - L - 1 = 0.$$

Solving by the quadratic formula yields two solutions, one of which is phi. More about the second solution in a moment.

Notice that the above plausibility argument did not use the values $F(1)$ and $F(2)$. Indeed, more generally, if you take any additive sequence (any sequence — no matter what the first two terms — in which the third term and beyond are obtained by adding the preceding two), one gets the same result: convergence to phi. The first two numbers need not even be whole numbers or positive. This, too, is easy to test in a program that you can write yourself.

Thus, if we extend the definition of black hole to require only that the iterates get closer and closer to one number, we have a black hole once again.

But there is still another black hole one can derive from this. Consider the function $f(x) = 1+1/x$ for nonzero real numbers x. I selected (or rather stumbled across) this function f because of the simple argument above that gave phi as a limit. Start with a seed number x and iterate function values $f(x), f(f(x))$, etc. One obtains convergence to phi — but still no appearance of the second number that is the solution to the quadratic equation above.

Is there a connection between the two solutions of the quadratic equation? First, each number is the negative reciprocal of the other. Each is also one minus the other. Second, had we formed the ratios $F(n)/F(n+1)$, we would have obtained the absolute value of the second solution instead of the first solution.

Third, note that our last function $f, f(x) = 1+1/x$, could not use an input of 0 because division by zero is undefined. The solution to the equation $f(x) = 1+1/x = 0$ is just $x = -1$. Thus, x may not equal -1 either.

We have not finished. We must now avoid $f(x) = -1$ (otherwise, $f(f(x)) = 0$ and then $f(f(f(x)))$ is undefined). Solving $1 + 1/x = -1$ gives $x = -1/2$. If we continue now working backward with function f's preimages we find, in succession, that we must similarly rule out $-\frac{2}{3}$, then $-\frac{3}{5}$, $-\frac{5}{8}$, $-\frac{8}{13}$, etc. Notice that these fractions are precisely the negatives of the ratios of consecutive Fibonacci numbers in the reverse order that we considered. All of these must be eliminated from the universe for f, along with 0. The limiting ratio, the second golden ratio, must also be eliminated from the universe.

In closing out this brief connection of Fibonacci numbers to the topic, I would be remiss if I did not follow my own advice on getting black holes by looking for fixed points: values x such that $f(x) = x$. For $f(x) = 1+1/x$, we obtain our two golden numbers. As things are set up, the number $1.618\ldots$ is an attractor with the black-hole property, while $-0.618\ldots$ is a repeller. All real inputs except zero, the numbers $-F(n-1)/F(n)$, and the second golden number lead to the attractor, our black hole called phi.

Unsolved Problems as Black Holes

Even unsolved problems sometimes fall into this black-hole scenario. Consider the Collatz Conjecture, dating back to the 1930s and still an open question (though sometimes also identified with the names of Hailstone, Ulam, and Syracuse). Start with a natural number. If odd, triple and add one. If even, take half. Keep iterating. Must you always reach 1?

If you start with 5, you get 16, then 8, then 4, then 2, then 1. Success! In fact, this problem has the paradoxical property that, although one of the hardest to settle definitively, is among the easiest to program (a few lines).

If you do reach 1 – and nobody has either proved you must, nor shown any example that doesn't – then you next get 4, then 2, then 1 again, a cycle that repeats ad infinitum. Hmm... a cycle of length three. We're interested now only in black holes, which really are cycles of length one. So, let's be creative and fix this up by modifying the problem.

Define the process instead by taking the starting number and breaking it down completely into factors that are odd and even. For instance, 84 is $2 \times 2 \times 3 \times 7$. Pick the largest odd factor. In the example that would be $3 \times 7 = 21$. (Just multiply all the odd prime factors together. The only non-odd prime is 2.) Now triple the largest odd factor and then add 1. This answer is the next iterate.

Now try some examples. You should find that you keep getting 4. Once you hit 4, you stay at 4, because the largest odd factor in 4 is 1, and $3 \times 1 + 1 = 4$. Anybody who proves the Collatz Conjecture will prove that my variation is a mathemagical black hole as well, and conversely.

Because this is so easy to program, I will omit a program here.

Close

Other examples of mathemagical black holes arise in the study of stochastic processes. Under certain conditions, iteration of matrix powers draws one to a result that represents long-term stability independent of an input vector. There are striking resemblances to convergence results in such disciplines as differential equations, too.

Quite apart from any utility, however, the teasers and problems here are intriguing in their own right. Moreover, the real value for me is in seeing the black-hole idea as the unifying theme of seemingly disparate recreations. This is an ongoing pursuit in my own newsletter, *Recreational and Eductional Computing,* in which we've had additional ones. I invite readers to send me other examples or correspondence:

Dr. Michael W. Ecker, Editor
Recreational and Educational Computing
909 Violet Terrace
Clarks Summit, Pennsylvania 18411-9206 USA

Happy Black-Hole hunting during your salute to Martin Gardner, whom we are proud to make a tiny claim to as our Senior Contributing Mathematical Editor.

Puzzles from Around the World

Richard I. Hess

Introduction

Most of the puzzles in this collection were presented in the Logigram, a company newsletter published at Logicon, Inc., where I worked for 27 years. The regular problem column appeared from 1984 to 1994 and was called "Puzzles from Around the World"; it consisted of problems from a number of sources: other problem columns, other solvers through word of mouth, embellishments or adaptations of such problems, or entirely new problems of my own creation. These problems are for the enjoyment of the solver and should be passed on to others for their enjoyment as well.

The problems are arranged in three general categories of *easy*, *medium*, and *hard*; the solution section gives an approach to answering each of the problems. A sources section provides information on the source of each problem as far as I know. I would be delighted to hear from anyone who can improve the solution approach, add information on the source of the problem, or offer more interesting problems for future enjoyment.

I thank the following people for providing problems and ideas that have contributed to the richness of the problems involved: Leon Bankoff, Brian Barwell, Nick Baxter, Laurie Brokenshire, James Dalgety, Clayton Dodge, Martin Gardner, Dieter Gebhardt, Allan Gottlieb, Yoshiyuki Kotani, Harry Nelson, Karl Scherer, David Singmaster, Naoaki Takashima, Dario Uri, Bob Wainwright, and, especially, Nob Yoshigahara. Source information begins on page 82.

Easy Problems

E1. A man has breakfast at his camp. He gets up and travels due North. After going 10 miles in a straight line he stops for lunch. After lunch he gets up and travels due North. After going 10 miles in a straight line he finds himself back at camp. Where on earth could he be?

E2. Find the rule for combining numbers as shown below (e. g., 25 and 9 combine to give 16) and use the rule to determine x.

63 9 38 33 32 12

88 25 16 18 15 x 5

E3. What number belongs at x in this sequence?

34 32 36 46 64 75 50 35 34

16 18 14 22 x 40 35 15 20 12

E4. It is approximately 2244.5 nautical miles from Los Angeles to Honolulu. A boat starts from being at rest in Los Angeles Harbor and proceeds at 1 knot per hour to Honolulu. How long does it take?

E5. You have containers that hold 15 pints, 10 pints, and 6 pints. The 15-pint container starts out full, and the other two start out empty: (15, 0, 0). Through transferring liquid among the containers, measure exactly two pints for yourself to drink and end up with 8 pints in the 10-pint container and 5 pints in the 6-pint container. Find the most efficient solution.

E6. You are at a lake and have two empty containers capable of holding exactly π (= 3.14159...) and e (= 2.7182818...) liters of liquid. How many transfers of liquid will it take you to get a volume of liquid in one container that is within one percent of exactly one liter?

E7. Calculate the expansion of this 26 term expression:

$$E = (x-a)(x-b)(x-c)(x-d)...(x-y)(x-z)$$

E8. Find a nine-digit number made up of 1, 2, 3, 4, 5, 6, 7, 8, 9 in some permutation such that when digits are removed one at a time from the right the number remaining is divisible in turn by 8, 7, 6, 5, 4, 3, 2 and 1.

E9. My regular racquetball opponent has a license plate whose three-digit part has the following property. Divide the number by 3, reverse the digits of the result, subtract 1 and you produce the original number. What is the number and what is the next greater number (possibly with more than three digits) having this property?

E10. Find nine single-digit numbers other than (1, 2, 3, ..., and 9) with a sum of 45 and a product of $9! = 362{,}880$.

E11. A knight is placed on an infinite checkerboard. If it cannot move to a square previously visited, how can you make it unable to move in as few moves as possible?

E12. Place the numbers from 1 to 15 in the 3×5 array so that each column has the same sum and each row has the same sum.

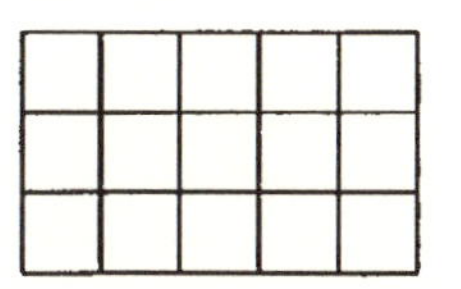

E13. The Bridge. Four men must cross a bridge. They all start on the same side and have 17 minutes to get across. It is night, and they need their one flashlight to guide them on any crossing. A maximum of two people can cross at one time. Each man walks at a different speed: A takes 1 minute to cross; B takes 2 minutes; C takes 5 minutes, and D takes 10 minutes. A pair must walk together at the rate of the slower man's pace. Can all four men cross the bridge? If so, how?

Try these other problems.

(a) There are six men with crossing times of 1, 3, 4, 6, 8, and 9 minutes and they must cross in 31 minutes.

(b) There are seven men with crossing times of 1, 2, 6, 7, 8, 9, and 10 minutes, and the bridge will hold up to three men at a time, and they must cross in 25 minutes.

E14. Divide the figure below into

(a) 4 congruent pieces and

(b) 3 congruent pieces.

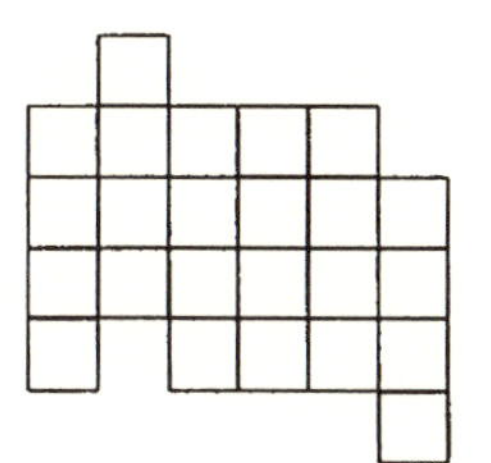

E15. Potato Curves. You are allowed to draw a closed path on the surface of each of the potatoes shown below. Can you draw the two paths so that they are identical to each other in three-dimensional space?

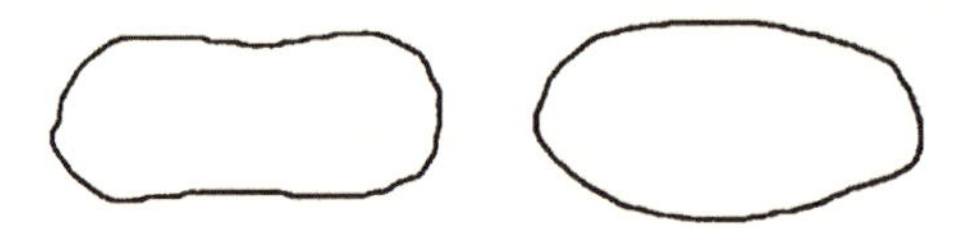

E16. Suppose a clock's second hand is exactly on a second mark and exactly 18 second marks ahead of the hour hand. What time is it?

E17. Shoelace Clock. You are given some matches, a shoelace, and a pair of scissors. The lace burns irregularly like a fuse and takes 60 minutes to burn from end to end. It has a symmetry property in that the burn rate a distance x from the left end is the same as the burn rate the same distance x from the right end. What is the minimum time interval you can measure?

Medium Problems

M1. You have a 37-pint container full of a refreshing drink. N thirsty customers arrive, one having an 11-pint container and another having a $2N$-pint container. How will you most efficiently measure out 1 pint of drink for each customer to drink in turn and end up with N pints in the 11-pint container and $37\text{-}2N$ pints in the 37-pint container if

(a) $N = 3$;
(b) $N = 5$?

M2. Three points have been chosen randomly from the vertices of a cube. What is the probability that they form (a) an acute triangle; (b) a right triangle?

M3. You were playing bridge as South and held 432 in spades, hearts, and diamonds. In clubs you held 5432. Despite your lack of power you took 6 tricks, making a 4-club contract. Produce the other hands and a line of play that allows this to occur.

M4. From England comes the series ... 35, 45, 60, x, 120, 180, 280, 450, 744, 1260, ... Find a simple continuous function to generate the series and compute the surprise answer for x.

M5. How many ways can four points be arranged in the plane so that the six distances between pairs of points take on only two different values?

M6. From the USSR we are asked to simplify $x = \sqrt[3]{2 + \frac{10}{3\sqrt{3}}} + \sqrt[3]{2 - \frac{10}{3\sqrt{3}}}$.

M7. Find all primes p such that $2^p + p^2$ is also prime. Prove there are no others.

M8. The river shown below is 10 feet wide and has a jog in it. You wish to cross from the south to the north side and have only two thin planks of length L and width 1 ft to help you get across. What is the least value for L that allows a successful plan for crossing the river?

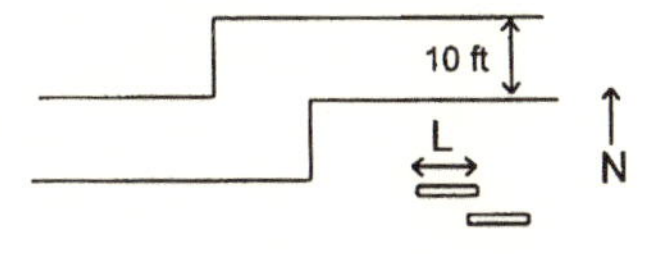

M9. A regular pentagon is drawn on ordinary graph paper. Show that no more than two of its vertices lie on grid points.

M10. 26 packages labeled A to Z are known to each weigh whole numbers of pounds in the range 1 to 26.

(a) Determine the weight of each package with a two-pan balance and four weights of your own design.
(b) Now do it with three weights.

M11. Music on the planet Alpha Lyra IV consists of only the notes A and B. Also, it never includes three repetitions of any sequence nor does the repetition BB ever occur. What is the longest Lyran musical composition?

M12. Many crypto doorknob locks use doors with five buttons numbered from 1 to 5. Legal combinations allow the buttons to be pushed in specific order either singly or in pairs without pushing any button more than once. Thus [(12), (34)] = [(21), (34)]; [(1), (3)]; and [(2), (13), (4)] are legal combinations while [(1), (14)]; [(134)]; and [(13), (14)] are not.

(a) How many legal combinations are there?
(b) If a sixth button were added, how many legal combinations would there be?

M13. Find the smallest prime number that contains each digit from 1 to 9 at least once.

M14. Dissect a square into similar rectangles with sides in the ratio of 2 to 1 such that no two rectangles are the same size. A solution with nine rectangles is known.

M15. Divide an equilateral triangle into three contiguous regions of identical shape if

(a) All three regions are the same size;
(b) all three regions are of different size;
(c) two of the regions are the same size and the third region is a different size.

M16. Dissect a square into similar right triangles with legs in the ratio of 2 to 1 such that no two triangles are the same size. A solution with eight triangles is known.

M17. You and two other people have numbers written on your foreheads. You are told that the three numbers are primes and that they form the sides of a triangle with prime perimeter. You see 5 and 7 on the other two people, both of whom state that they cannot deduce the number on their own foreheads. What number is written on your forehead?

M18. A snail starts crawling from one end along a uniformly stretched elastic band. It crawls at a rate of 1 foot per minute. The band is initially 100 feet long and is instantaneously and uniformly stretched an additional 100 feet at the end of each minute. The snail maintains his grip on the band during the instant of each stretch. At what points in time is the snail (a) closest to the far end of the band, and (b) farthest from the far end of the band?

M19. An ant crawls along the surface of a $1 \times 1 \times 2$ "dicube" shown below.

(a) If the ant starts at point A, which point is the greatest distance away? (It is not B.)
(b) What are the two points farthest apart from each other on the surface of the dicube? (The distance, *d*, between these points is greater than 3.01).

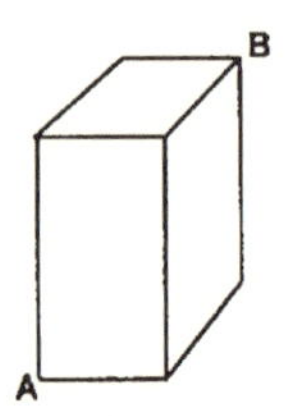

M20. Humpty Dumpty. "You don't like arithmetic, child? I don't very much," said Humpty Dumpty.

"But I thought you were good at sums," said Alice.

"So I am," said Humpty Dumpty. "Good at sums, oh certainly. But what has that got to do with liking them? When I qualified as a Good Egg – many, many years ago, that was – I got a better mark in arithmetic than any of the others who qualified. Not that that's saying a lot. None of us did as well in arithmetic as in any other subject."

"How many subjects were there?" said Alice, interested.

"Ah!" said Humpty Dumpty, "I must think. The number of subjects was one third of the number of marks obtainable in any one subject. And I ought to mention that in no two subjects did I get the same mark, and that is also true of the other Good Eggs who qualified."

"But you haven't told me ... ," began Alice.

"I know I haven't told you how many marks in all one had to obtain to qualify. Well, I'll tell you now. It was a number equal to four times the maximum obtainable in one subject. And we all just managed to qualify."

"But how many ... ," said Alice.

"I'm coming to that," said Humpty Dumpty. "How many of us were there? Well, when I tell you that no two of us obtained the same assortment of marks – a thing which was only just possible – you'll be well on the way to the answer. But to make it as easy as I can for you, I'll put it another way. The number of other Good Eggs who qualified when I did, multiplied by the number of subjects (I've told you about that already), gives a product equal to half the number of marks obtained by each Good Egg. And now you can find out all you want to know."

He composed himself for a nap. Alice was almost in tears. "I can't," she said, "do any of it. Isn't it differential equations, or something I've never learned?"

Humpty Dumpty opened one eye. "Don't be a fool, child," he said crossly. "Anyone ought to be able to do it who is able to count on five fingers."

What was Humpty Dumpty's mark in Arithmetic?

Hard Problems

H1. Similar Triangles. The figure following this problem shows a 30-60-90 triangle divided into four triangles of the same shape. How many ways can you find to do this? (Nineteen solutions are known.)

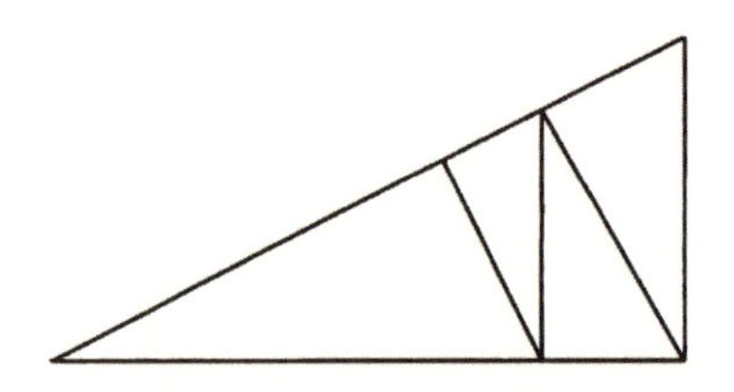

H2. "To reward you for killing the dragon," the Queen said to Sir George, "I grant you the land you can walk around in a day." She pointed to a pile of wooden stakes. "Take some of these stakes with you," she continued. "Pound them into the ground along your way, and be back at your starting point in 24 hours. All the land in the convex hull of your stakes will be yours." (The Queen had read a little mathematics.)

Assume that it takes Sir George 1 minute to pound a stake and that he walks at a constant speed between stakes. How many stakes should he take with him to get as much land as possible?

H3. An irrational punch centered on point P in the plane removes all points from the plane that are an irrational distance from P. What is the least number of irrational punches needed to eliminate all points of the plane?

H4. Imagine a rubber band stretched around the world and over a building as shown below. Given that the width of the building is 125 ft and the rubber band must stretch an extra 10 cm to accommodate the building, how tall is the building? (Use 20,902,851 ft for the radius of the earth.)

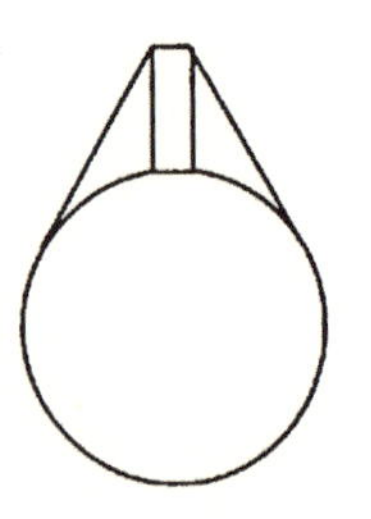

H5. A billiard ball with a small black dot, P, on the exact top is resting on the horizontal plane. It rolls without slipping or twisting so that its contact point with the plane follows a horizontal circle of radius equal to that of the ball. Where is the black dot when the ball returns to its initial resting place?

H6. Locate m points in the plane (perhaps with some exactly on top of others) so that each of them is a unit distance from exactly n of the others for $(m,n) =$

(a) (3, 2),
(b) (7, 4),
(c) (11, 6),
(d) (9, 4),
(e) (12, 4),
(f) (8, 3),
(g) (12, 5).

H7. A good approximation of π using just two digits is $\pi = 3.1$.

(a) Find the best approximation using two digits of your choice. You may use $+, -, \times, /$, exponents, decimal points, and parentheses. No square roots or other functions are allowed.
(b) The same as part (a) except square roots are allowed.
(c) Show that $1/\pi$ can be approximated with arbitrary accuracy using any two digits if the rules of part (b) are followed.

H8. (a) What region inside a unit square has the greatest ratio of area to perimeter?

(b) What volume inside a unit cube has the greatest ratio of volume to surface area?

H9. As shown below it is easy to place $2n$ unit diameter circles in a $2 \times n$ rectangle. What is the smallest value of n for which you can fit $2n + 1$ such circles into a $2 \times n$ rectangle?

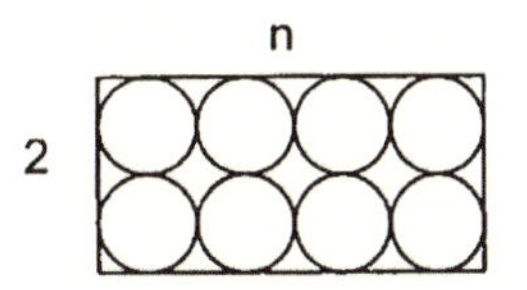

H10. You are given two pyramids $SABCD$ and $TABCD$. The altitudes of their eight triangular faces, taken from vertices S and T, are all equal to 1. Prove or disprove that line ST is perpendicular to plane $ABCD$.

H11. Tennis Paradox. Two evenly matched tennis players are playing a tiebreak set. The server in any game wins each point with a fixed probability p, where $0 < p < 1$. For what values of p and score situations during a set can the player ahead according to the score have less than a 50% chance of winning the set?

H12. My uncle's ritual for dressing each morning except Sunday includes a trip to the sock drawer, where he (1) picks out three socks at random, then (2) wears any matching pair and returns the odd sock to the drawer or (3) returns the three socks to the drawer if he has no matching pair and repeats steps (1) and (3) until he completes step (2). The drawer starts with 16 socks each Monday morning (eight blue, six black, two brown) and ends up with four socks each Saturday evening.

(a) On which day of the week does he average the longest time to dress?
(b) On which day of the week is he least likely to get a pair from the first three socks chosen?

H13. The hostess at her 20th wedding anniversary party tells you that the youngest of her three children likes her to pose this problem, and proceeds to explain: "I normally ask guests to determine the ages of my three children, given the sum and product of their ages. Since Smith missed the problem tonight and Jones missed it at the party two years ago, I'll let you off the hook." Your response is "No need to tell me more, their ages are . . . "

H14. Minimum Cutting Length. What is the minimum cut-length needed to divide

(a) a unit-sided equilateral triangle into four parts of equal area?
(b) a unit square, into five parts of equal area?
(c) an equilateral triangle into five parts of equal area?

H15. A number may be approximated by a/b where a and b are integers. Define the goodness of fit of a/b to x as $g(a, b, x) = a|bx - a|$. An approximation to π is 355/113, with $g(355, 113, \pi) = 0.0107$.

(a) Find the minimum a and b so that $g(a, b, e) < 0.0107 \ldots$, where $e = 2.71828\ldots$.
b) Find the minimum c and d so that $g(c, d, \pi) < 0.0107\ldots$.

H16. (a) The figure shows two Pythagorean triangles with a common side where three of the five side-lengths are prime numbers. Find other such examples.

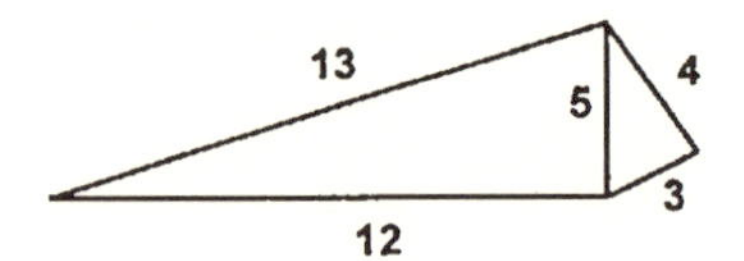

(b) Can a third Pythagorean triangle be abutted such that 4 of the 7 lengths are primes?

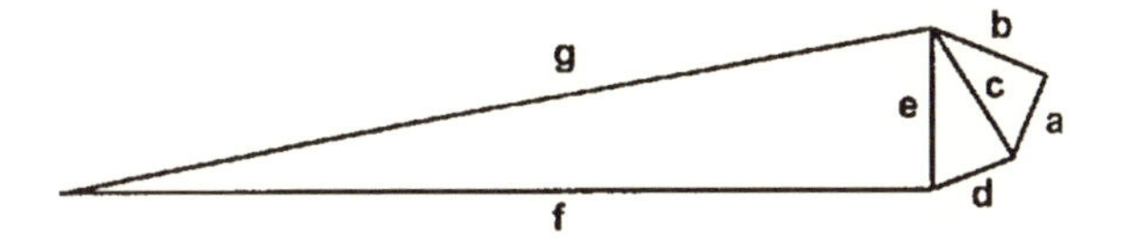

H17. The inhabitants of Lyra III recognize special years when their age is of the form $a = p^2q$, where p and q are different prime numbers. The first few such special years are 12, 18 and 20. On Lyra III one is a student until reaching a special year immediately following a special year; one then becomes a master until reaching a year that is the third in a row of consecutive special years; finally one becomes a sage until death, which occurs in a special year that is the fourth in a row of consecutive special years.

(a) When does one become a master?
(b) When does one become a sage?
(c) How long do the Lyrans live?
(d) Do five or six special years ever occur consecutively?

H18. The planet Alpha Lyra IV is an oblate spheroid. Its axis of rotation coincides with the spheroid's small axis, just as one observes for Sol III. It's internal structure, however, is unique, being made of a number of coaxial right circular cylindrical layers, each of homogeneous composition. The common axis of these layers is the planet's rotational axis. The outermost of these layers is nearly pure krypton, and the next inner layer is an anhydrous fromage. Cosmic, Inc., is contemplating mining the outer layer, and the company's financial planners have found that the venture will be sound if there are more than one million cubic spandrals of krypton in the outer layer. (A spandral is the Lyran unit of length.)

Unfortunately, the little that is known about the krypton layer was received via sub-etherial communication from a Venerian pioneer immediately prior to its demise at the hands (some would say wattles) of a frumious snatcherband. The pioneer reported with typical Venerian obscurantism that the ratio of the volume of the smallest sphere that could contain the planet to the volume of the largest sphere that could be contained within the planet is 1.331 to 1. It (the Venerians are sexless) also reported that the straight line distance between it and its copod was 120 spandrals. (A copod corresponds roughly to something between a sibling and a rootshoot.) The reporting Venerian was mildly comforted because the distance to its copod was the minimum possible distance between the two. By nature the Venerians can only survive at the krypton/fromage boundary and by tragic mistake the two copods had landed on disconnected branches of the curve of intersection of the krypton/fromage boundary and the planetary surface.

Should Cosmic, Inc., undertake the mining venture?

H19. Taurus, a moon of a-Lyra IV (hieronymous), was occupied by a race of knife-makers eons ago. Before they were wiped out by a permeous accretion of Pfister-gas, they dug a series of channels in the satellite surface. A curious feature of these channels is that each is a complete and perfect circle, lying along the intersection of a plane with the satellite's surface. An even more curious fact is that Taurus is a torus (doughnut shape).

Five students of Taurus and its ancient culture were discussing their field work one day when the following facts were brought to light:

- The first student had dug the entire length of one of the channels in search of ancient daggers. He found nothing but the fact that the length of the channel was 30π spandrals.
- The second student was very tired from his work. He had dug the entire length of a longer channel but never crossed the path of the first dagger digger.
- The third student had explored a channel 50π spandrals in length, crossing the channel of the haggard dagger digger.
- The fourth student, a rather lazy fellow (a laggard dagger digger?), had merely walked the 60π spandral length of another channel, swearing at the difficulties he had in crossing the channel of the haggard dagger digger.
- The fifth student, a rather boastful sort, was also tired because he had thoroughly dug the entire length of the largest possible channel.

How long was the channel which the braggart haggard dagger digger dug?

H20. Define

$$F(n) = 2n + \frac{2}{3} - e^n \sum_{k=0}^{n-1} (k-n)^k e^{-k}/k!,$$

where $n = 1, 2, 3, \ldots$ and $e = 2.7182818\ldots$

(a) Prove that $F(n)$ goes to zero as n goes to infinity.
(b) Find $F(1000)$ to three significant figures.
(c) Find the smallest m such that the magnitude of $F(m)$ is less than the magnitude of $F(m+1)$.

Solutions to Easy Problems

E1. Anywhere within 10 miles of the north pole.

E2. At first glance, it appears that the rule might be subtraction, and $x = 17$. But this is not right because 18 is not the difference of 38 and 16. Instead, the rule is that two numbers combine to give a number that is the sum of the digits of the two numbers. Thus $x = 11$.

E3. In each case the numbers combine by summing the products of their digits. Thus $x = 4 \times 6 + 2 \times 2 = 28$.

E4. The trick here is that 1 knot = 1 nautical mile per hour, so 1 knot per hour is a constant acceleration of 1 nmi per hour per hour.
$2244.5 = \frac{at^2}{2}$ gives $t = 67$ hours.

E5. 15 moves: (15, 0, 0), (9, 0, 6), (9, 6, 0), (3, 6, 6), (3, 10, 2), (3, 10, 0), (3, 4, 6), (7, 0, 6), (7, 6, 0), (1, 6, 6), (1, 10, 2), (11, 0, 2), (11, 2, 0), (5, 2, 6), (5, 8, 0), (0, 8, 5).

E6. The goal is to find the smallest integer values of a and b so that

$$n = 2(a + b - 1)$$

is a minimum and $0.99 < ae - b\pi < 1.01$ or $0.99 < a\pi - be < 1.01$.
By numerical search we find

$$73\pi - 84e = 1.00059 \text{ and } 57e - 49\pi = 1.004024$$

are the smallest values satisfying the above equations. Thus $a = 57$, $b = 49$, and $n = 210$ transfers is the minimum.

E7. One of the terms is $(x - x)$, so $E = 0$.

E8. The number is 381,654,729.

E9. The license number is 741; the next greater number is 7,425,741.

E10. (1, 2, 4, 4, 4, 5, 7, 9, 9).

E11. The figure below shows how to trap the knight after 15 moves.

				1				
	3							
			2		0			
		4				14		
5				15				13
		6				12		
			8		10			
	7						11	
				9				

E12. See the array below.

1	14	2	12	11
8	6	9	7	10
15	4	13	5	3

E13. Yes; *A* and *B* go across, *A* comes back; *C* and *D* go across, *B* comes back; *A* and *B* go across.

(a) 1 and 3 go across, 1 comes back; 8 and 9 go across, 3 comes back; 1 and 6 go across, 1 comes back; 1 and 4 go across, 1 comes back; 1 and 3 go across.

(b) 1 and 2 go across, 1 comes back; 8, 9, and 10 go across, 2 comes back; 1, 6,and 7 go across, 1 comes back; 1 and 2 go across.

E14. The solutions are shown in the following figures.

E15. Imagine intersecting the potatoes with each other. The path of their intersecting surfaces is a desired path.

E16. If the hour hand is exactly on a second mark then the second hand will always be on the 12. For the second hand to be 18 second marks ahead of the hour hand the hour hand must be at the 42nd second mark, and the time is 8:24.

E17. Cut the lace in half, producing pieces A and B. Burn A from both ends, noting the point that burns last. Cut B at that corresponding point, producing pieces C and D. At the same time, start burning C from both ends and D from one end. When C is consumed, light the other end of D. In 3.75 minutes, D will finish burning. This is the shortest time interval that can be measured.

Solutions to Medium Problems

M1. (a) 32 moves; (37, 0, 0), (31, 0, 6), (31, 6, 0), (25, 6, 6), (25, 11, 1), (36, 0, 1), (36, 0, 0), (30, 0, 6), (30, 6, 0), (24, 6, 6), (24, 11, 1), (35, 0, 1), (35, 0, 0), (29, 0, 6), (29, 6, 0), (23, 6, 6), (23, 11, 1), (23, 11, 0), (23, 5, 6), (29, 5, 0), (29, 0, 5), (18, 11, 5), (18, 10, 6), (24, 10, 0), (24, 4, 6), (30, 4, 0), (30, 0, 4), (19, 11, 4), (19, 9, 6), (25, 9, 0), (25, 3, 6), (31, 3, 0)

(b) 40 moves; (37, 0, 0), (26, 11, 0), (26, 1, 10), (26, 0, 10), (36, 0, 0), (25, 11, 0), (25, 1, 10), (25, 0, 10), (35, 0, 0), (24, 11, 0), (24, 1, 10), (34, 1, 0), (34, 0, 0), (23, 11, 0), (23, 1, 10), (33, 1, 0), (33, 0, 0), (22, 11, 0), (22, 1, 10), (22, 0, 10), (11, 11, 10), (21, 11, 0), (21, 1, 10), (31, 1, 0), (31, 0, 1), (20, 11, 1), (20, 2, 10), (30, 2, 0), (30, 0, 2), (19, 11, 2), (19, 3, 10), (29, 3, 0), (29, 0, 3), (18, 11, 3), (18, 4, 10), (28, 4, 0), (28, 0, 4), (17, 11, 4), (17, 5, 10), (27, 5, 0).

M2. There are $7 \times 6/2 = 21$ choices using vertex 1. See the figure. The probability of an acute triangle is $3/21$ or $1/7$; the probability of a right triangle is $18/21$ or $6/7$, as seen from the list below. $A = 135, 137, 157$. $R =$ 123, 124, 125, 126, 127, 128, 134, 136, 138, 145, 146, 147, 148, 156, 158, 167, 168, 178.

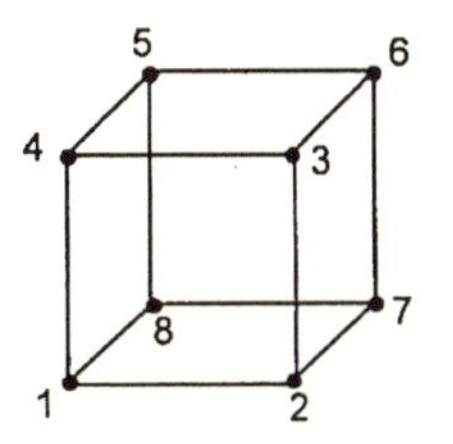

M3. The hand below does the job if the play goes as follows. The first five tricks are alternate spade and heart ruffs by North and West, with East underruffing North each time. North wins the sixth trick with the 10 of diamonds, East playing the 9. North next leads a heart which West trumps with the queen. The next four tricks are won by South with the 5432 of clubs; North and East discard all their diamonds on these tricks. The final two tricks are won by South's long diamonds.

```
                          S -
                          H 98765
                          D 108765
                          C J97
S AKQJ1098765                                S -
H -                                          H AKQJ10
D -                       S 432              D AKQJ9
C AKQ                     H 432              C 1086
                          D 432
                          C 5432
```

M4. The series can be expressed by a simple continuous function:

$$F(n) = 120(2^n - 1)/n, \text{for } n \neq 0.$$

To get x, we take the limit of $F(n)$ as n goes to 0.

$$x = 120 \ln 2 = 83.17766\ldots.$$

M5. There are six ways as shown below.

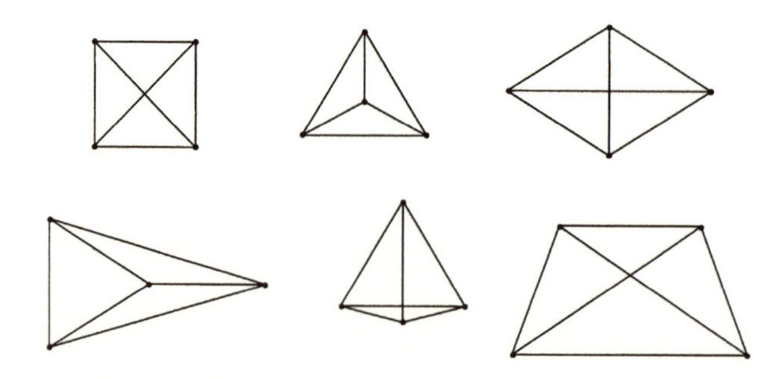

M6. $x^3 - 2x - 4 = 0$. The only real root is $x = 2$.

M7. For $p = 3$ we get the prime 17. For $p > 3$ we have $p = 6k + 1$ or $p = 6k - 1$. Then $2^p = 3n - 1$ and $p^2 = 3m + 1$, so that $2^p + p^2$ is always a multiple of 3 for $p > 3$.

M8. In the figure below the oblique lines each have a length of $(20\sqrt{2})/3 = 9.42809$ feet. If these lines are interpreted as the diagonals of the planks then the planks can be marginally longer than $L = \sqrt{791}/3 = 9.3749074$ feet.

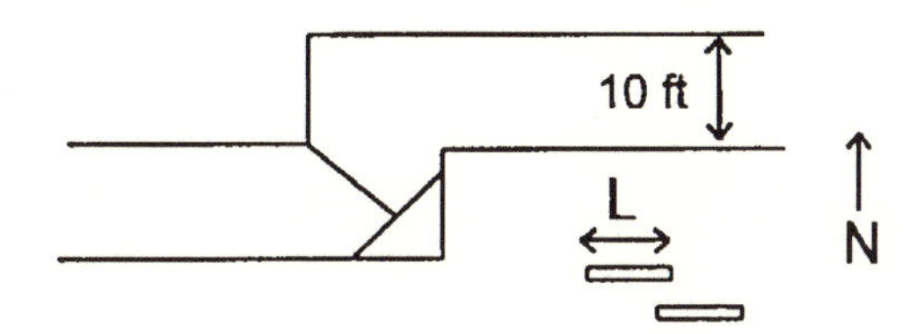

M9. Assume that three vertices do fall on grid points. The triangle they form will always include a vertex with an angle of 36 degrees as shown below.

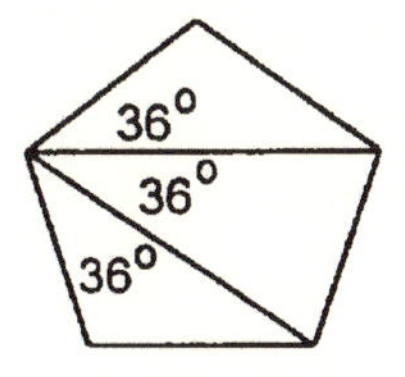

Place that vertex at the origin and suppose the other two vertices of the triangle are at grid points (a, b) and (c, d). Then

$$\cos(36^\circ) = \frac{1+\sqrt{5}}{4} = \frac{ac+bd}{\sqrt{(a^2+b^2)(c^2+d^2)}} \rightarrow 3+\sqrt{5} = \frac{8(ac+bd)^2}{(a^2+b^2)(c^2+d^2)}$$

But $\sqrt{5}$ is irrational and therefore cannot be the ratio of two integers. This contradiction proves that three vertices of a regular pentagon cannot lie on grid points.

M10. (a) Weights of 1, 3, 9, and 27 do the job.

(b) Weights of 2, 6, and 18 do the job. Note that a 1-pound package is the only one that won't balance or lift a 2-pound weight.

M11. The longest musical composition without three consecutive repetitions of any sequence are AABABAABABAABAAB and its reverse.

M12. (a) For 5 buttons there are the following types of combinations where S is a single button pushed and P is a pair of buttons pushed: S, 2S, 3S, 4S, 5S, P, PS, P2S, P3S, 2P, 2PS. Corresponding to each of these types the number of combinations is

$$\frac{5!}{4!}+\frac{5!}{3!}+\frac{5!}{2!}+\frac{5!}{1!}+\frac{5!}{0!}+\frac{5!}{2\cdot 3!}+\frac{2\cdot 5!}{2\cdot 2!}+$$

$$\frac{3\cdot 5!}{2\cdot 1!}+\frac{4\cdot 5!}{2\cdot 0!}+\frac{5!}{2\cdot 2!}+\frac{3\cdot 5!}{2\cdot 2!} = 935$$

(b) For 6 buttons the types are: S, 2S, 3S, 4S, 5S, 6S, P, PS, P2S, P3S, P4S, 2P, 2PS, 2P2S, and 3P. The number of combinations is

$$\frac{6!}{5!}+\frac{6!}{4!}+\frac{6!}{3!}+\frac{6!}{2!}+\frac{6!}{1!}+\frac{6!}{0!}+\frac{6!}{2\cdot 4!}+\frac{2\cdot 6!}{2\cdot 3!}+\frac{3\cdot 6!}{2\cdot 2!}+$$

$$\frac{4\cdot 6!}{2\cdot 1!}+\frac{5\cdot 6!}{2\cdot 0!}+\frac{6!}{2\cdot 2\cdot 2}+\frac{3\cdot 6!}{2\cdot 2}+\frac{6\cdot 6!}{2\cdot 2}+\frac{6!}{2\cdot 2\cdot 2} = 7671.$$

M13. 1,123,465,789.

M14. A nine-rectangle solution is shown below. Each number is the length of the shorter side of the rectangle.

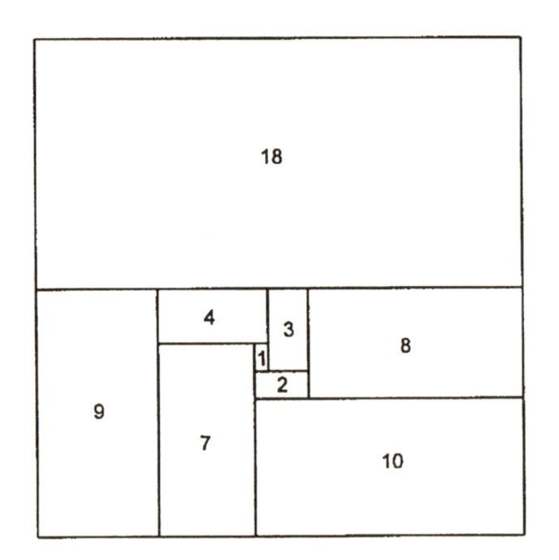

M15. Shown below.

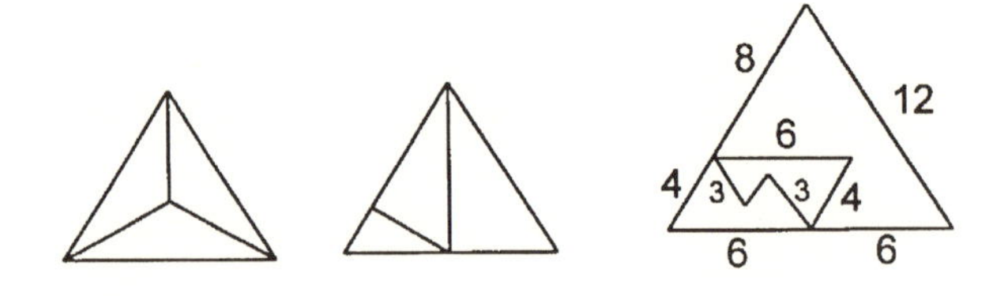

M16. An eight-triangle solution is shown below.

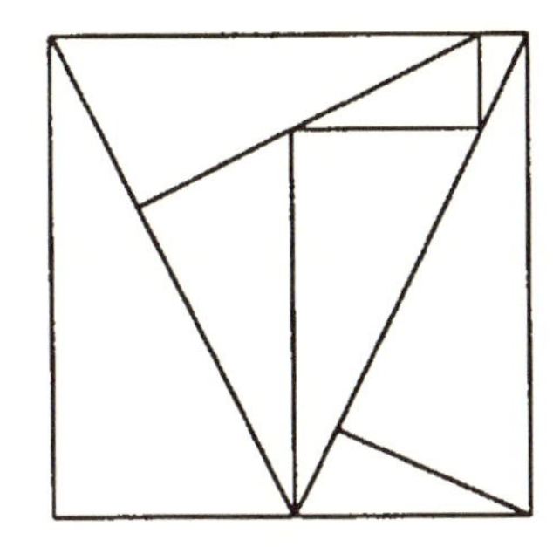

M17. (1) The three possibilities are that you have 5, 7 or 11 on your forehead.

(2) If you have a 5, then person A with a 7 sees (5, 5) and concludes that he must have 3 or 7. But if he has 3 then he reasons that person B sees (5, 3) and would know his number is 5 or 3. B can eliminate 3 because anyone seeing (3, 3) would immediately know he had 5. Since B doesn't know his number, A would conclude that he has a 7. Since A doesn't draw this conclusion you know you don't have a 5.

(3) If you have a 7 then person B with a 5 sees (7, 7) and concludes he has 3 or 5. But if he has 3 then he reasons that person A with 7 sees (7, 3) and would know his number is 7. Since A doesn't know his number, then B would conclude he has a 5. Since B doesn't draw this conclusion you know you don't have a 7. Therefore you have 11 on your forehead.

M18. (a) During the nth minute the band is $100n$ long and the snail travels a fraction of the band equal to $1/100n$. The snail eventually arrives at the far end when $1+\frac{1}{2}+\frac{1}{3}+\ldots+\frac{1}{t} = 100$. That happens for $t = 1.509269 \times 10^{43}$ minutes.

(b) When the snail has traversed 99 % of the band, the 100-foot stretch causes the far end to move 1 foot farther from the snail. This 1 foot is overcome by the snail's normal progress during that minute. This happens when $1 + \frac{1}{2} + \frac{1}{3} + \cdots + \frac{1}{u} = 99$, or $u = 5.5522899 \times 10^{42} = t/e$.

M19. (a) Unfold the box and examine shortest path distances from A to various points. The farthest is at point P one quarter of the way along the diagonal on the 1×1 face from B.

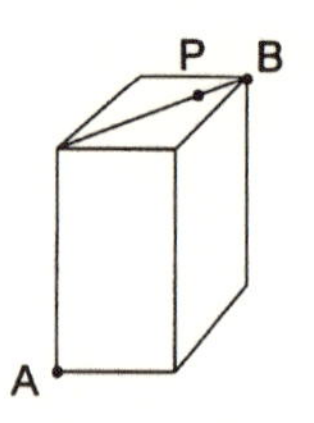

(b) The two points are along the diagonal near P and its opposite point on the 1×1 face that includes A. The points are $(\sqrt{3} - 1)/2$ of the way along each diagonal for a distance of 3.0119 … .

M20. There are 7 Good Eggs. There are 5 subjects with 15 marks possible in each. The scores for the Good Eggs are (15, 14, 13, 12, 6), (15, 14, 12, 11, 7), (15, 14, 13, 10, 8), (15, 14, 12, 11, 8), (15, 14, 12, 10, 9), (15, 13, 12, 11, 9) and (14, 13, 12, 11, 10). Humpty Dumpty got a 10 in arithmetic.

Solutions to Hard Problems

H1. Shown below. The number in each triangle indicates its area relative to the other triangles in the figure.

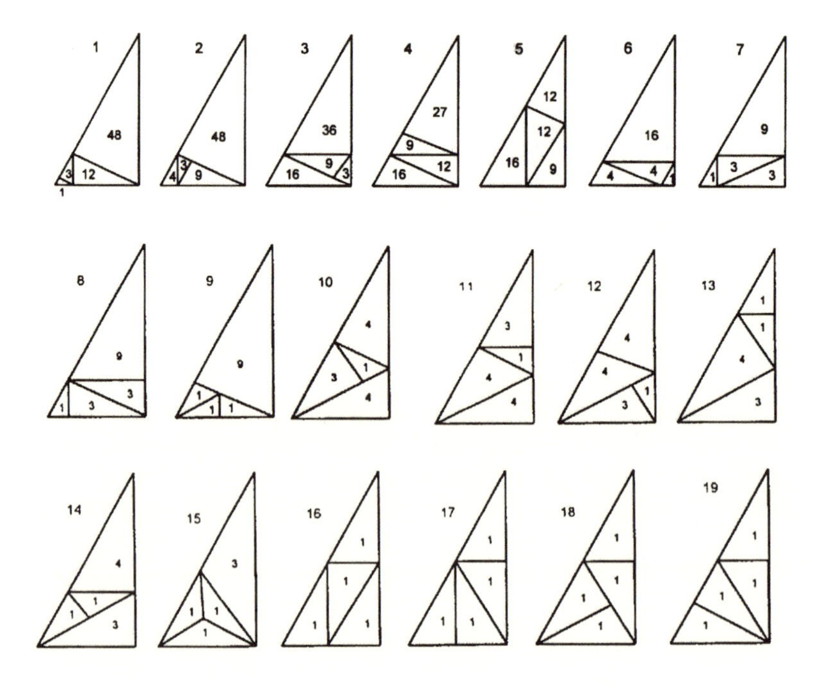

H2. Imagine a regular polygon of n sides after Sir George's trip. Its area is $A = nsh/2$, where s is the side length and h is the distance from the center of the polygon to the center of a side: $(2h)/s = \cot(180/n)$. If Sir George travels at speed v then $s/v = (1440 - n)/n$ where 1440 is the number of minutes in a day. To maximize the area is to find the n that maximizes

$$A = v^2 \cot(\frac{180}{n})(1440 - n)^2/(4n).$$

The area is largest for $n = 17$, with $A =$ 159,300.1968 v^2.

H3. Three equally spaced punches in a straight line do the job if the square of the spacing is irrational.

H4. Let w = width of the building; h = height of the building; r = the earth's radius; and d = stretch in the band due to the building. Then

$$2\tan B = \frac{w}{r+h};$$

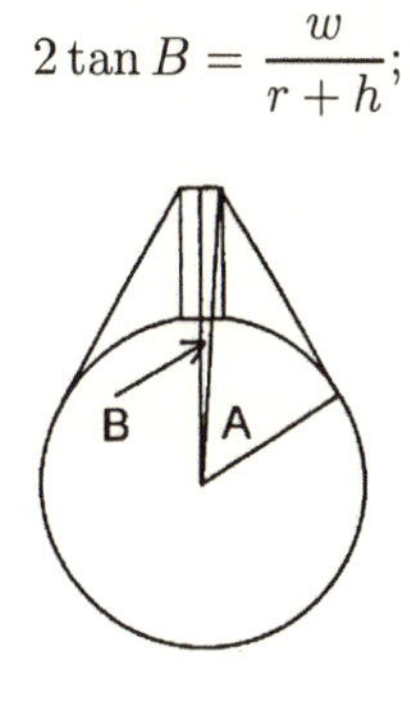

$$r \tan A = \sqrt{2rh + h^2 + \frac{w^2}{4}};$$

$$d = 2r\tan A + w - 2r(A+B).$$

These equations can be iterated to produce $h = 85.763515\ldots$ ft.

H5. Consider the ball rotating about the axis AC without slipping, as shown below. The cone, BCB' rolls along the plane. The distance from B to the axis AC is $\sqrt{0.5}$ of the distance BC. Thus B will be in contact with the horizontal plane when $\sqrt{0.5}$ rotations about the axis AC have occurred. Point P will return to the top again as well. For a full rotation of the ball on the circle the point P will execute $\sqrt{2}$ rotations about the axis. Relative to the center point of the ball, the point P will have coordinates $P = -r\sin\theta/\sqrt{2}, r(\cos\theta - 1)/2, r(\cos\theta + 1)/2$, where $\theta = \sqrt{8}\pi$. For $r = 1, P = (-0.3629497, -0.9291081, 0.0708919)$; its initial position was (0, 0, 1).

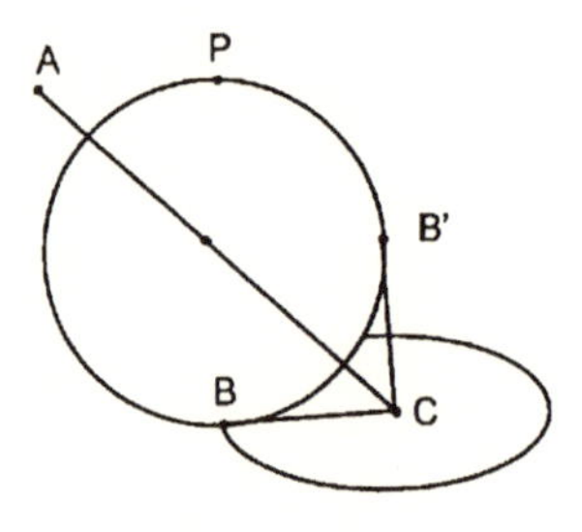

H6. The solutions are shown below.

(a)

(b) $\theta \neq 0°, 60°, 180°, 240°, \text{or } 300°$.

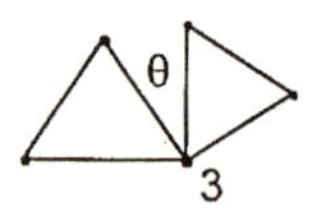

(c) $\theta \neq 0°, 60°, \text{or} 180°$.

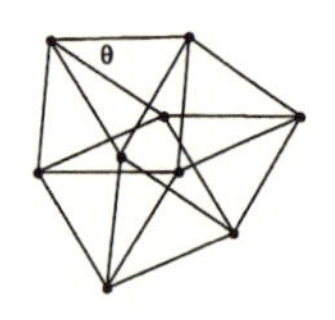

(d) θ, ϕ and $\theta + \phi + 60° \neq 0°, 60°, 180°, 240°$, or $300°$.

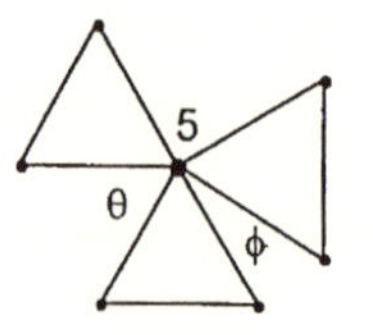

(e)

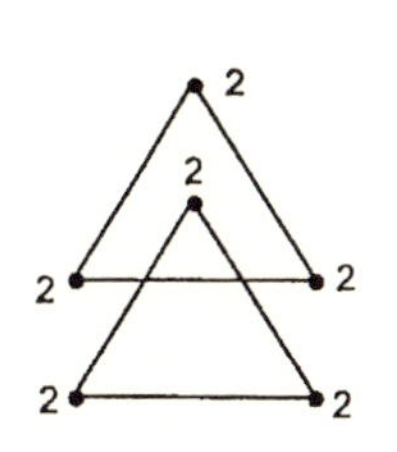

(f) θ, ϕ and $\theta + \phi \neq 0°, 60°$, or $120°$.
$\cos(2\phi) + \cos(2\theta) \neq 2\cos\phi + 2\cos(\theta) - \cos(\theta - \phi)$.

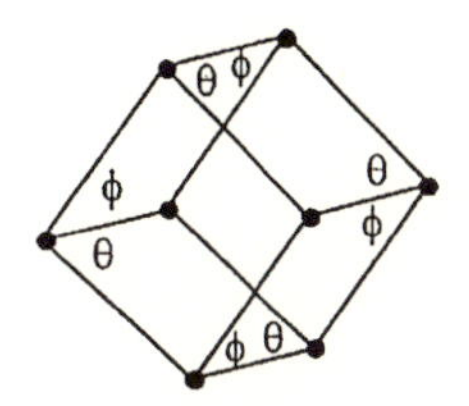

(g)

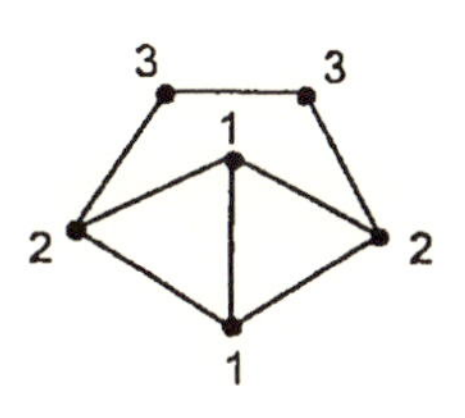

H7. (a) $.1^{-.5} = 3.162276\ldots$

(b) $\dfrac{\sqrt{\sqrt{\sqrt{\sqrt{.9}}}}}{\sqrt{.1}} = 3.141522\ldots.$

(c) Define $a(n) = \sqrt{\cdots\sqrt{a}\cdots}$ where there are n square roots. Clearly $c = a(n) - b(m)$ can be made arbitrarily close to 0. For the right k, $c(k)$ can be put in the interval $0.25 - 0.5$. Pick a and b as any digits other than 0. There are an infinite number of choices of m and n followed by the appropriate k to make an infinite density of $c(k)$'s in the $0.25 - 0.5$ interval, which contains $1/\pi$.

H8. (a) Imagine an expanding soap bubble in the unit square. Its form will take on the shape of four quarter circles of radius r at the corners as shown in the figure. The area of such a shape is $A = 1 + (\pi - 4)r^2$.

The perimeter of the shape is $P = 2\pi r + 4 - 8r$. The ratio A/P takes on a maximum when $r = 1/(2+\sqrt{\pi}) = 0.265079\ldots$. The maximum ratio $A/P = r = 0.265079\ldots$. Note that the ratio A/P is 0.25 for the entire square or a circle inscribed in the square.

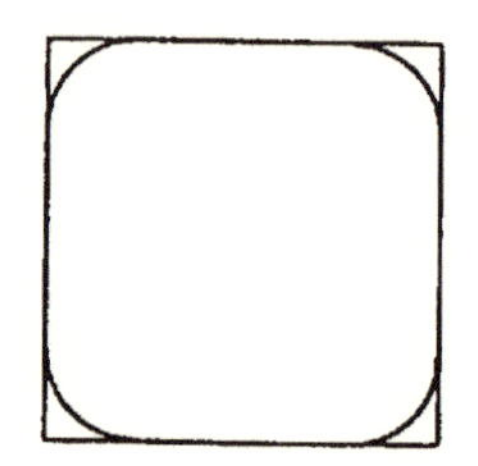

(b) The corresponding problem of maximizing the ratio of volume to surface area within the unit cube remains unsolved.

H9. The circles need to be packed as shown below. For $n = 164$ there is just enough room for 329 circles. There are 7 circles on each end with 105 sets of 3 circles in the middle. The smallest rectangle found to date containing 329 circles has 13 circles on each side of 101 sets of 3 and measures $2 \times 163.9973967\ldots$.

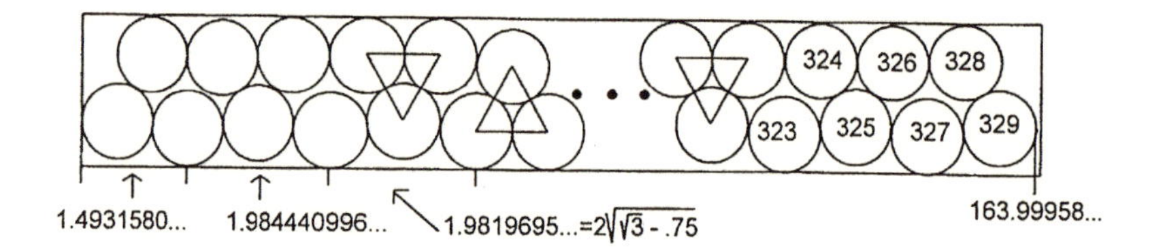

H10. A view of the pyramid looking down from S perpendicular to the plane $ABCD$ will look like the figure below.

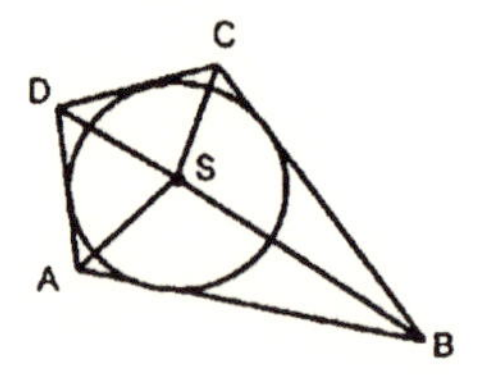

The point S may be raised above the plane to give unit altitudes if the inscribing circle has radius < 1. Now consider two such circles of different sizes as shown. Each has a radius < 1. They define the four tangent lines drawn to produce $ABCD$ as shown. Raise S and T appropriately above

the plane so that all altitudes are unit. Clearly ST is not perpendicular to the plane.

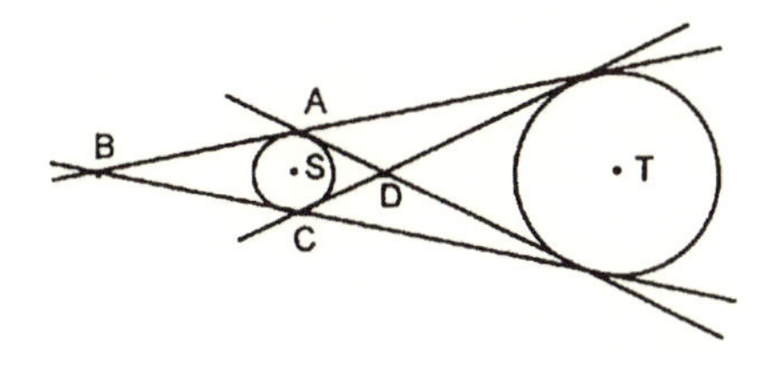

H11. Let $P(A - B)$ be the probability of the server winning the game when the server has A points and the receiver has B points. Let $p =$ the probability of the server winning a point, and let $q = 1 - p$.
$P(40 - 40) = pP(40 - 30) + qP(30 - 40)$;
$P(40 - 30) = p + qP(40, 40)$;
$P(30 - 40) = pP(40 - 40)$.
This leads to
$P(40 - 40) = \frac{p^2}{p^2+q^2}$;
$P(40 - 30) = p + \frac{p^2q}{p^2+q^2}$.
Similarly one derives
$P(40 - 15) = p + pq + \frac{p^2q^2}{p^2+q^2}$;
$P(30 - 15) = p^2(1 + q) + \frac{p^2q(pq+1)}{p^2+q^2}$;
$P(0 - 0) = \frac{p^4(1-16q^4)}{p^4-q^4}$.
At a score of n games to n ($n = 0$ to 5), $P(0 - 0) > P(30 - 15)$ when $8p^2 - 4p > 3$, which leads to $p > 0.911437827$. At n games to n, $P(0 - 0) > P(40 - 30)$ when $8p^3 - 4p^2 - 2p > 1$, which leads to $p > 0.919643377607$.

H12. A computer program was written to calculate the probabilities shown in the table below. My uncle is expected to take the longest time to dress on Saturday and on Friday he is least likely to get a pair from the first three socks chosen.

Day	Average Selections Required	Pairing probability
Monday	1.2069	0.8286
Tuesday	1.2270	0.8163
Wednesday	1.2511	0.8027
Thursday	1.2801	0.7892
Friday	1.3074	0.7805
Saturday	1.3586	0 .7849

H13. We look for cases where (1) the oldest child is under 20, (2) the younger two children have different ages, and (3) there is a product and sum that give rise to ambiguity for the ages both this year and two years ago. There is only one set of ages that accomplishies this: (5, 6, 16). The product and sum could be achieved by (4, 8, 15), which must have been guessed by Smith. The product and sum two years ago could be achieved by (2, 7, 12), which must have been guessed by Jones two years ago.

H14. (a) Length = 1.342181807 as shown.
(b) Length = 2.50211293 as shown.
(c) Unknown.

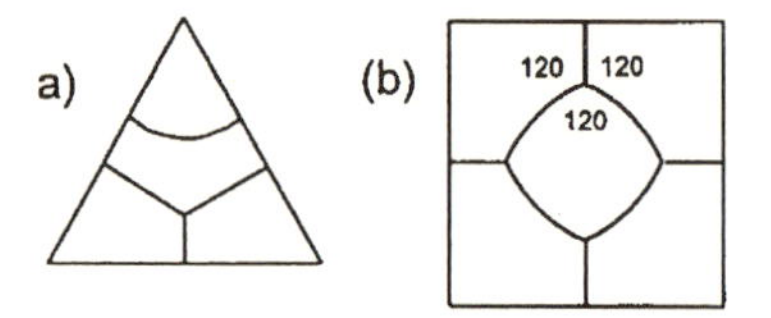

H15. Using a multiprecision continued fraction program, we get

(a) $a = 7.187824 \times 10^{288}$, $b = 2.69442527 \times 10^{288}$, $g = 0.010659$.
(b) $c = 4.744902 \times 10^{154}$, $d = 1.51034933 \times 10^{154}$, $g = 0.007190$.

H16. (a) $a =$ prime, $b = (a^2 - 1)/2$, $c = (a^2 + 1)/2 =$ prime, $d = (c^2 - 1)/2$, $e = (c^2 + 1)/2 =$ prime.

Solutions occur for $a = 3, 11, 19, 59, 271, 349, 521, 929, 1031, 1051,\ 1171, \ldots$.
(b) $a = 271$; $b = 36,720$; $c = 36,721$; $d = 674,215,920$; $e = d + 1$; $f = 227,283,554,064,939,120$; $g = f + 1$.

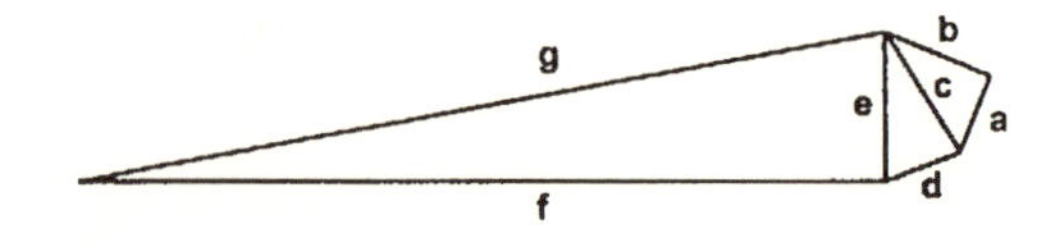

H17. (a) 45 years.
(b) 605 years.
(c) 17,042,641,444 years.
(d) Yes; a=10,093,613,546,512,321 is the first of 5 in a row.
$a = 49p1, a + 1 = 2 * p * p, a + 2 = 9 * p2, a + 3 = 4 * p3, a + 4 = 25 * p4.$

Six in a row is impossible because three consecutive even special years cannot occur.

H18. The venture should be undertaken since the volume in cubic spandrals can be determined using calculus to be

$$V = \frac{4\pi}{3}(1.331)^{2/3}(60)^3 = 1,094,782.208.$$

It is noteworthy that the volume doesn't depend on the polar or equilateral radii of Alpha Lyra IV.

H19. The figure below shows a cross-section of the torus.

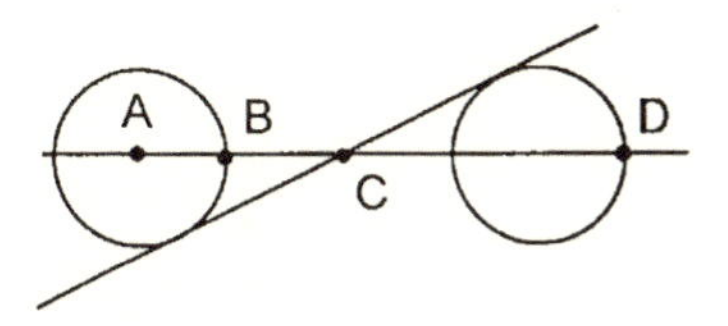

Let $r = AB$ and $R = CD$. There are three classes of channels that can be dug on its surface.

(a) It is clear that many channels with radius r are possible.
(b) Channels with radii between $R - 2r$ and R are possible.
(c) A less well-known third type of circular channel with radius $R - r$ is possible, which is the intersection of the plane shown by the slanting line and the torus.

From the descriptions of the first two students they must have dug channels of type (b). The third and fourth students must have explored channels of type (a) and (c) in some order. One case gives $R - r = 25$ and $r = 30$; the other case gives $R - r = 30$ and $r = 25$. The first case is not possible because $R - 2r < 0$. Thus the second case applies and $2\pi R = 110\pi$.

H20. (1) Define

$$I(n,x) = \sum_{k=0}^{n} \frac{(k-n)^k e^{(k-n)x}}{k!} = \sum_{k=0}^{n} \frac{1}{k!}\frac{d^k}{dx^k}e^{(k-n)x}$$

Clearly $F(n) = 2n + \frac{2}{3} - I(n,-1)$.

(2) Since $e^{(k-n)z}$ is an entire function in the complex plane, it follows from Cauchy's Theorem that

$$F(n) = 2n + \frac{2}{3} - \sum_{k=0}^{n} \frac{1}{2\pi i} \int_C dz \frac{e^{(k-n)z}}{(1+z)^{k+1}},$$

where C must enclose $z = -1$ and will be taken to be $|z| = 2$.

(3) The summation over k can be carried out explicitly to give

$$F(n) = 2n + \frac{2}{3} + \frac{1}{2\pi i} \int_C dz \frac{e^{-(n+1)z}}{1-(1+z)e^{-z}} - \frac{1}{2\pi i} \int_C \frac{dz}{[1-(1+z)e^{-z}](1+z)^{n+1}}$$

(4) Define $q(z) = 1/[1-(1+z)e^{-z}]$. The only pole of $q(z)e^{-(n+1)z}$ within C is a double pole at $z = 0$. Thus $z^2 q(z) e^{-(n+1)z}$ is regular within C, and by Cauchy's Theorem we have

$$\frac{1}{2\pi i} \int_C dz q(z) e^{-(n+1)z} = \frac{d}{dz} \left[z^2 q(z) e^{-(n+1)z} \right]_{z=0} = -(2n + \frac{2}{3})$$

Thus

$$F(n) = -\frac{1}{2\pi i} \int_C \frac{dz q(z)}{(1+z)^{n+1}}$$

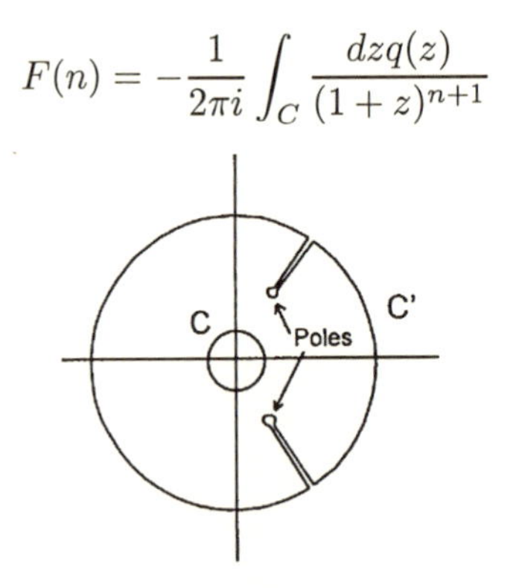

To evaluate $F(n)$, we expand and deform C to a contour C' like that in the figure, which still avoids enclosing the poles of $q(z)$. As C' gets arbitrarily large it still encloses the same poles as C. Thus

$$\begin{aligned} F(n) &= -\frac{1}{2\pi i} \int_{C'} \frac{dz q(z)}{(1+z)^{n+1}} \\ &= -\frac{1}{2\pi i} \int_{|z|=R} \frac{dz q(z)}{(1+z)^{n+1}} + \frac{1}{2\pi i} \sum_{k=1}^{k=\infty} \int_{C_k} \frac{dz q(z)}{(1+z)^{n+1}}, \end{aligned}$$

where C_k is the counterclockwise path about the kth pole of $q(z)$, not including its pole at $z = 0$.
(5) On $|z| = R$ for sufficiently large R, $q(z)$ is bounded and the integral on the path tends to zero as R tends to infinity. Thus $F(n)$ is the sum of integrals on small circle paths about the poles of $q(z)$. A pole of $q(z)$ occurs at $1 + z = e^z$, where $z = x + iy$. The resulting equations in x and y are $\sin y = ye^{(1-y\cot y)}$ and $x = \ln(y/\sin y)$. Since the equation in y is even, we can look for poles in the upper half-plane only and reflect each one into the lower half-plane. Call the kth such pole in the upper half-plane $z_k = x_k + iy_k$. The table below gives numerical values of the first few. For larger k define $\phi_k = (2k + 1/2)\pi$. Then y_k approximately equals $\phi_k - [1 + \ln(\phi_k)]/\phi_k$.

k	x_k	y_k
1	2.0888430	7.4614892
2	2.6406814	13.8790560
3	3.0262969	20.2238350
4	3.2916783	26.5432385
5	3.5012690	32.8505482
6	3.6745053	39.1512074
7	3.8221528	45.4473849

(6) The contribution to $F(n)$ from pole k is the limit as z approaches z_k of $(z - z_k)q(z)/(1 + z)^{n+1}$, which reduces to

$$(\cos(n+1)y_k - i\sin(n+1)y_k)\,\frac{(1 + x_k + iy_k)(x_k - iy_k)}{D},$$

where $D = (x_k^2 + y_k^2)e^{(n+1)x_k}$.
Combining the poles in the upper and lower half-planes and some rearrangement finally produces

$$F(n) = \sum_{k=1}^{\infty} A_k e^{-nx_k} \sin(ny_k - \theta_k),$$

where

$$A_k = \frac{-2\sec(y_k + \theta_k)e^{-x_k}y_k}{x_k^2 + y_k^2}$$

$$\tan(y_k + \theta_k) = \frac{x_k^2 + x_k + y_k^2}{y_k}.$$

(a) Since $x_k > 0$ for all k, e^{-nx_k} goes to zero as n goes to infinity.

(b) For $n > 10$, $F(n)$ is dominated by the first pole and its reflected pole. Thus $F(1000)$ is very nearly $A_1 e^{-1000x_1} \sin(1000y_1 - \theta_1)$, giving $F(1000) = -1.14698 \times 10^{-909}$.

(c) From observing the behavior of $\sin(ny_1 - \theta_1)$ one can determine the smallest m where the magnitude of $F(m)$ is less than the magnitude of $F(m+1)$. It is at $m = 800$.

$$F(800) = 5.06175 \times 10^{-728}, F(801) = -5.12298 \times 10^{-728}$$

Sources

All problems written by the author unless otherwise indicated.
E1. From David Singmaster (private communication).
E2. From Nobuyuki Yoshigahara (private communication).
E3. From Nobuyuki Yoshigahara (private communication).
E4. From David Singmaster (private communication).
E5. Problem 3, "Puzzle Corner," in MIT's *Technology Review,* edited by Alan Gottlieb, May-June 1990, p. MIT 55, ©1990. All rights reserved. Reprinted with permission.
E7. Unknown.
E8. From Nobuyuki Yoshigahara (private communication).
E10. From Nobuyuki Yoshigahara (private communication).
E11. From Yoshiyuki Kotani (private communication).
E12. Unknown.
E13. Unknown; (a) and (b) original problems.
E14. From Nobuyuki Yoshigahara (private communication).
E15. From Dieter Gebhardt (private communication).
E16. Unknown.
M2. Unknown.
M3. Modification of Problem 1, "Puzzle Corner," in MIT's *Technology Review,* edited by Allan Gottlieb, October 1987, p. MIT 59, ©1987. Submitted by Lawrence Kells. All rights reserved. Reprinted with permission.
M4. Problem 1, "Puzzle Corner," in MIT's *Technology Review,* edited by Alan Gottlieb, October 1992, p. MIT 55, ©1992. From David Singmaster (attributed to Roger Penrose; private communication). All rights reserved. Reprinted with permission.
M5. Problem 3a, "Puzzle Section," *Pi Mu Epsilon Journal,* Vol. 8, No. 3, page 178, ©1985. Submitted by S. J. Einhorn and I. J. Schoenberg. All rights reserved. Reprinted with permission.
M6. Problem 7 from the 1980 Leningrad High School Olympiad, "Olympiad Corner," *Crux Mathematicorum,* Vol. 9, No. 10, p. 302, ©1983. All rights reserved. Reprinted with permission of the Canadian Mathematical Society.

M7. Problem 944, *Crux Mathematicorum,* Vol. 10, No. 5, p. 155, ©1984. Submitted by Cops of Ottawa. All rights reserved. Reprinted with permission of the Canadian Mathematical Society.
M11. From Yoshiyuki Kotani (private communication).
M14. From Robert T. Wainwright (private communication).
M15. From Karl Scherer (private communication).
M16. From Karl Scherer (private communication).
M17. Problem 4, "Puzzle Corner," in MIT's *Technology Review,* edited by Alan Gottlieb, October 1988, p. MIT 53, ©1988. All rights reserved. Reprinted with permission.
M18. (a) Unknown.
M19. (a) From Yoshiyuki Kotani (private communication); (b) original problem.
M20. From Robert T. Wainwright (private communication).
H1. From Robert T. Wainwright (private communication).
H2. Problem 1289, *Crux Mathematicorum,* Vol. 13, No. 9, p. 290, ©1987. Submitted by Carl Sutter. All rights reserved. Reprinted with permission of the Canadian Mathematical Society.
H3. Unknown.
H5. Problem 3, "Puzzle Corner," MIT's *Technology Review,* edited by Allan Gottlieb, May-June 1994, p. MIT 55, ©1994. Submitted by Bob High. All rights reserved. Reprinted with permission.
H6. Part (d): Problem 7, "Puzzle Section," *Pi Mu Epsilon Journal,* Vol. 8, No. 8, p. 526, ©1988. A special case of a problem by P. Erdos and G. Purdy. All rights reserved. Reprinted with permission.
H8. Part (a): Problem 870, *Crux Mathematicorum,* Vol. 12, No. 7, p. 180, ©1986. Submitted by Sydney Kravitz. All rights reserved. Reprinted with permission of the Canadian Mathematical Society.

Part (b): Problem 1225, *Crux Mathematicorum,* Vol. 13, No. 3, p. 86, ©1987. Submitted by David Singmaster. All rights reserved. Reprinted with permission.
H9. First published as Problem 3, "Puzzle Corner," in MIT's *Technology Review,* edited by Allan Gottlieb, January 1989, p. MIT 53, ©1989, and then as Department Problem 860, *Pi Mu Epsilon Journal,* Vol. 10, No. 2, p. 142, ©1995. Submitted by Nobuyuki Yoshigahara and J. Akiyama. All rights reserved. Reprinted with permission.
H10. Problem 1995, *Crux Mathematicorum,* Vol. 20, No. 10, p. 285, ©1994. Submitted by Jerzy Bednarczuk. All rights reserved. Reprinted with permission of the Canadian Mathematical Society.
H11. Problem 2, "Puzzle Corner," in MIT's *Technology Review,* edited by Allan Gottlieb, February-March 1997, p. MIT 47, ©1997. Submitted by Joe Shipman. All rights reserved. Reprinted with permission.
H12. Problem 1194, *Crux Mathematicorum,* Vol. 12, No. 10, 282, ©1986. All rights reserved. Reprinted with permission of the Canadian Mathematical Society.
H14. From Robert T. Wainwright (private communication).

H17. Problem 1231, *Crux Mathematicorum,* Vol. 13, No. 4, p. 118, ©1987. All rights reserved. Reprinted with permission of the Canadian Mathematical Society.

H18. Unknown.

H19. Unknown.

H20. Part (a): Problem 1190, *Crux Mathematicorum,* Vol. 12, No. 9, p. 242, ©1986. Submitted by Richard I. Hess from Albert Latter (private communication). All rights reserved. Reprinted with permission of the Canadian Mathematical Society.

O'Beirne's Hexiamond

Richard K. Guy

Tom O'Beirne is not as celebrated a puzzler as he deserves to be, particularly on this side of the Atlantic. When I began to write this article, I was sure that I would find references to the Hexiamond in Martin Gardner's column, but I haven't found one yet. Nor is it mentioned in O'Beirne's own book [4]. But it does appear in his column in the *New Scientist.* Maybe the only other places where it has appeared in print are Berlekamp et al. [1] and the not very accessible reference Guy [2].

It must have been in 1959 that O'Beirne noticed that, among the shapes that can be formed by adjoining six equilateral triangles, five had reflexive symmetry, while seven did not. Martin [3] uses O'Beirne's names for the shapes but does not distinguish between reflections so his problems only involve 12 shapes. If we count reflections as different, then there are 19 shapes (Figure 1). One of these is the regular hexagon, which can be surrounded by six more hexagons, and then by twelve more, giving a figure (Figure 2) having the same total area as the 19 shapes. Question: Will the 19 shapes cover the figure? It took O'Beirne some months to discover that the answer to the question is "Yes!" Figure 3 was discovered in November 1959.

O'Beirne thought that the result would be more pleasing if the Hexagon were in the center, and in January 1960 he found the solution shown in Figure 6. In the interim he had found solutions with the Hexagon in two other of its seven possible positions (Figures 4 and 5).

It was soon after this that O'Beirne visited the Guy family in London. He showed us many remarkable puzzles, but the one that grabbed us the most was the Hexiamond, and several copies had to be manufactured, since everyone wanted to try it at once. No one went to bed for about 48 hours. The next solution in my collection is Figure 7 by Mike Guy (March 1960).

We became adept at finding new solutions based on the old. Remove pieces $\bar{1}$, $\bar{3}$, 6, and 8 from Figure 3 and replace them in a different way. Or try using 0, 1, 3, 5, 6, and $\bar{8}$. Rearrange pieces 1, $\bar{1}$, 3, 5, and $\bar{8}$ in Figure 5; and $\bar{3}$, $\bar{6}$, 7, and $\bar{9}$ in Figure 7. It soon became necessary to devise a classification scheme, since it was not easy to decide whether a solution was new or not.

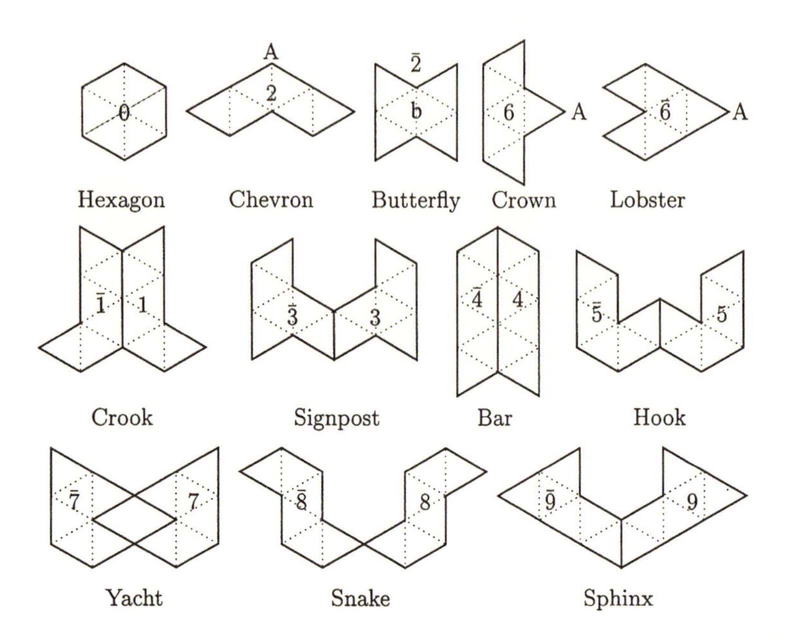

Figure 1. The nineteen Hexiamond pieces. The symmetries of the Hexagon are those of the dihedral group D6. The Butterfly has the same symmetries as a rectangle: Its position is described by that of its body. The Chevron, Crown, and Lobster each have a single symmetry of reflection, and each is positioned by its apex. Pieces 4 and 8 have only rotational symmetry and exist in enantiomorphous pairs, as do pieces 1, 3, 5, 7, and 9, which have no symmetry.

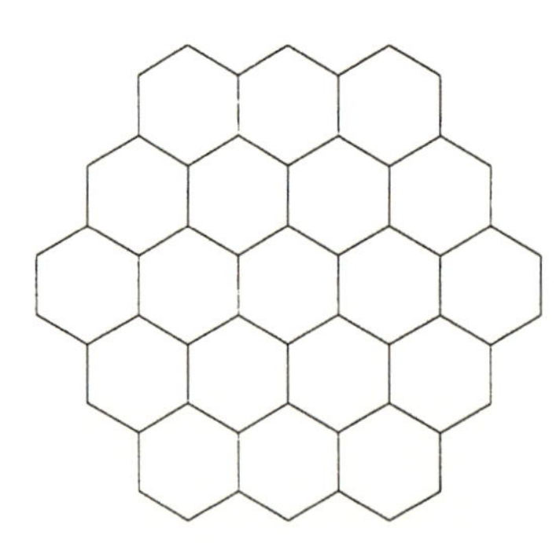

Figure 2. Hexiamond board.

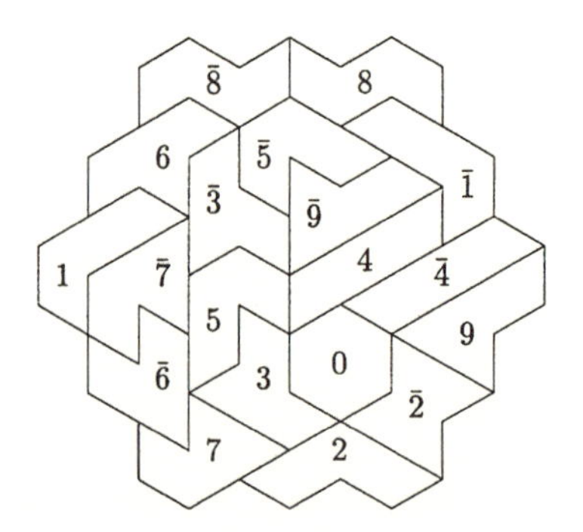

Figure 3. O'Beirne's first solution (November 30, 1959).

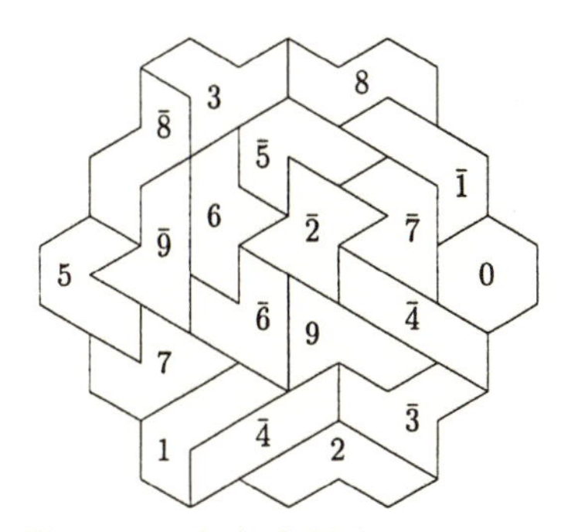

Figure 4. $hA_3f_4K_2l_5$ 60-01-05

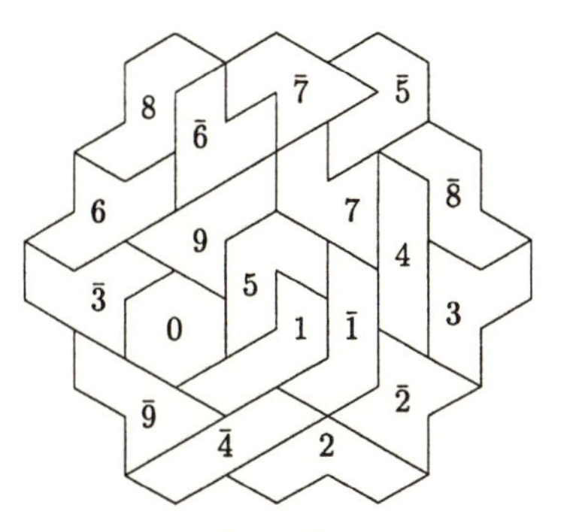

Figure 5. $hD_8p_4F_xd_7$ 60-01-10

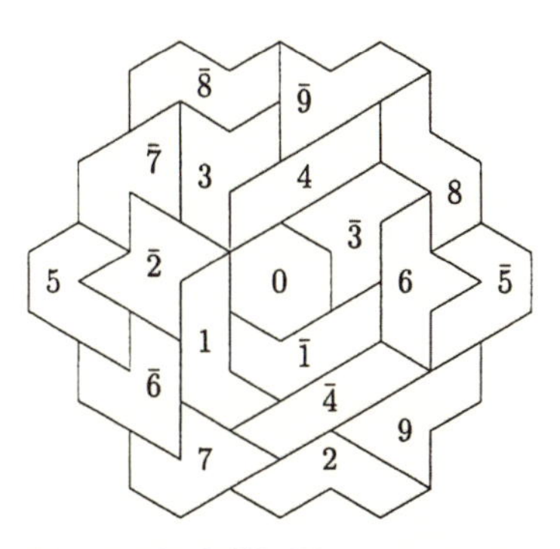

Figure 6. $hGh_xU_3a_5$ 60-01-21

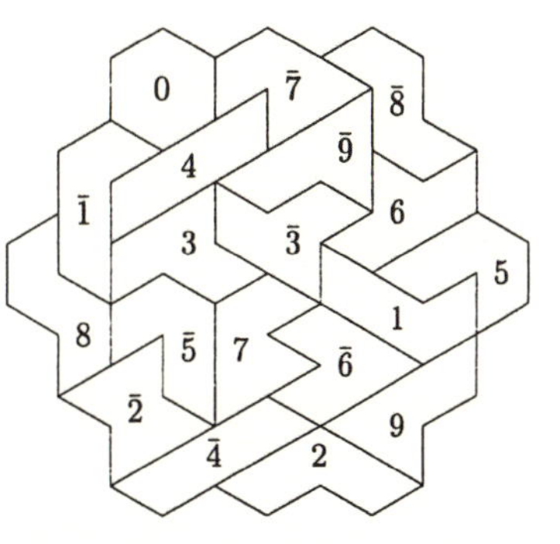

Figure 7. $hA_\varepsilon i_8B_\varepsilon i_3$ 60-03-27

O'Beirne had already suggested a numbering of the pieces from 1 to 19. By subtracting 10 from the Hexagon, and subtracting each of the labels 11 to 19 from 20 we arrive at the labeling in Figure 1. Notice that 0 has the greatest symmetry, the multiples of 4 have rotational symmetry through 180 degrees, and the other even numbers have reflective symmetries. Pieces with odd-numbered labels have no symmetry and, together with those that have rotational symmetry only, exist in enantiomorphous pairs. We don't know how O'Beirne decided whether to give a piece a number less than or greater than 10, but our mnemonics are as follows. Negative (bar sinister?) labels are given to the Bar, $\bar{4}$, a parallelogram drawn in the opposite way from the usual textbook fashion, to the Snake, $\bar{8}$, which appears to be turning to the left, to the Crook, $\bar{1}$, with its hook on the left, to the Signpost, $\bar{3}$, with its pointer on the left, to the Hook, $\bar{5}$, with its handle on the left, to the Yacht, $\bar{7}$, sailing to the left, and the Sphinx, $\bar{9}$, whose head is on the left.

We classify solutions with a string of five labels, giving the positions of the Chevron, Hexagon, Butterfly, Crown, and Lobster. This doesn't completely describe the solution, but gives enough detail to enable comparisons to be made quickly. We won't count a solution as different if it is just a rotation or a reflection, so first rotate the board so that the Chevron, piece number 2, is pointing upward. Then, if the Chevron is on the left-hand side of the board, reflect the board left to right in order to bring it onto the right half.

Read off the lowercase letter at the apex, *A*, of the Chevron from Figure 8. Note that the letters *a* and *n* are missing; it is possible to place the Chevron in such positions, but it's clear that they can't occur in a solution. More than three-quarters of the positions found so far have the Chevron in position *h*. (The exact figure is 89.5%.)

Next use Figure 9 to describe the position of the Hexagon, piece number 0. This will be a capital letter, *A* to *G*, depending on how far it is from the center. If the Chevron is central (positions *b* to *g* in Figure 8), reflect the board if necessary to bring the Hexagon into the right half. Then append a subscript $0, 1, \ldots, 9, x, \varepsilon$ indicating its position on the clock. *A* and *E* have only odd subscripts, *B, D, F* only even ones, and *G*, the center, is unique and requires no subscript. The subscripts on *C* are about $\frac{1}{2}$ more than their clock hour.

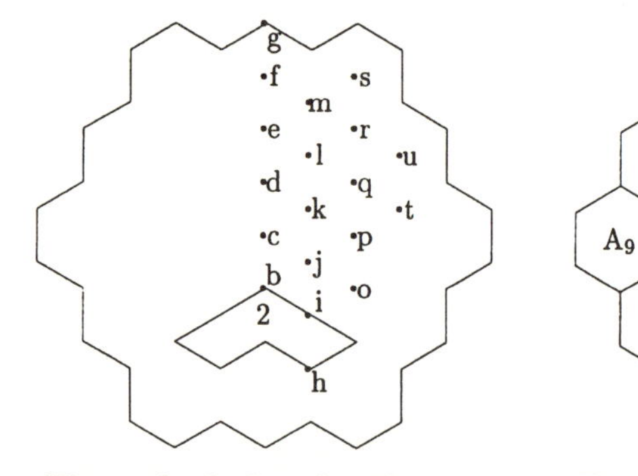

Figure 8. Coding the Chevron.

Figure 9. Coding the Hexagon.

The position of the Butterfly, piece number $\bar{2}$, is that of its "body," the edge that bisects it. This is indicated by a lowercase letter, shown in Figure 10, together with an even subscript, 0, 2, 4, 6, 8, or x, the side of the board it is nearest to. The subscripts *a, e, l, s,* and *t* are omitted, since placing the Butterfly there does not allow a solution to be completed; *f* and *m* are equidistant from opposite sides of the board, and are given only 0, 2 or 4 for a subscript. If both Chevron and Hexagon are symmetrically placed, reflect the board if necessary to bring the Butterfly into the right half. If all

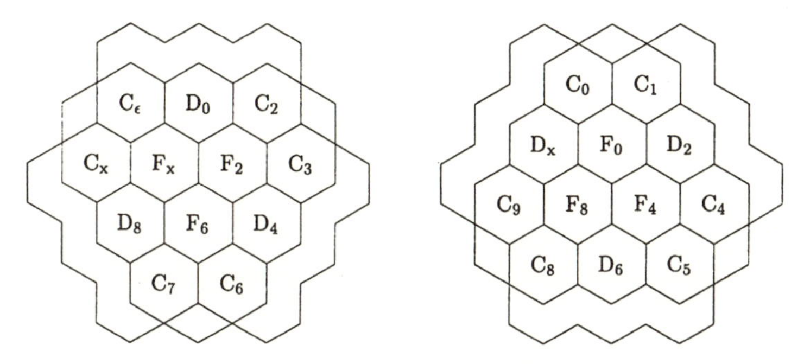

Figure 9. Continued: Coding the Hexagon's position.

three are symmetrically placed, as in the first solution in Figure 14, reflect if necessary to make the apex of the Crown point toward the right half of the board.

The position of the Crown, piece number 6, is given by the capital letter at its apex, *A*, in Figure 11. The subscript is even or odd according to whether the Crown is to the left or right of the axis of symmetry of the board in the direction in which its own axis of symmetry is pointed. Positions *S* and *T* are symmetrical and carry even subscripts. Positions *C, F, M, Q, R* are omitted, as they don't allow legal solutions.

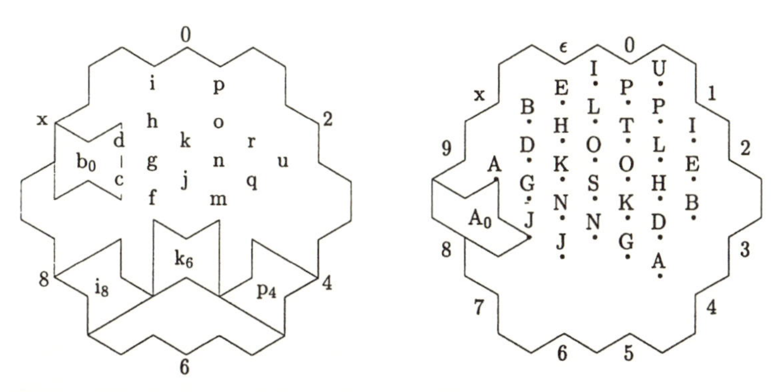

Figure 10. Coding the Butterfly.

Figure 11. Coding the Crown.

The Lobster, piece number $\bar{6}$, is located by a lowercase letter in Figure 12. Except for *a*, these are in the same positions as the capital letters used for the Crown. Again the subscript is even or odd according to whether the piece is to the left or right of the axis of symmetry of the board in the

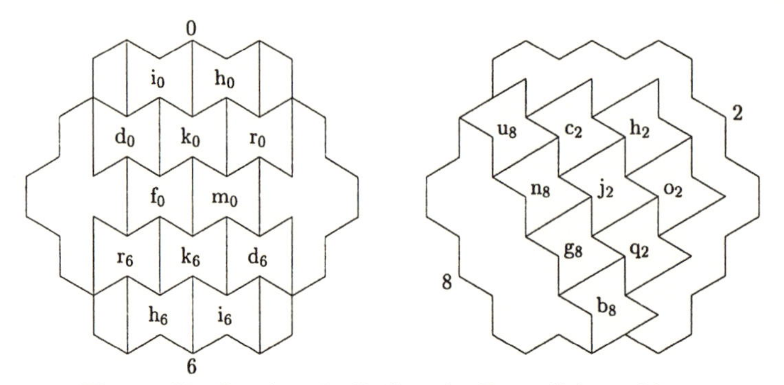

Figure 10. Continued: Coding the Butterfly's position.

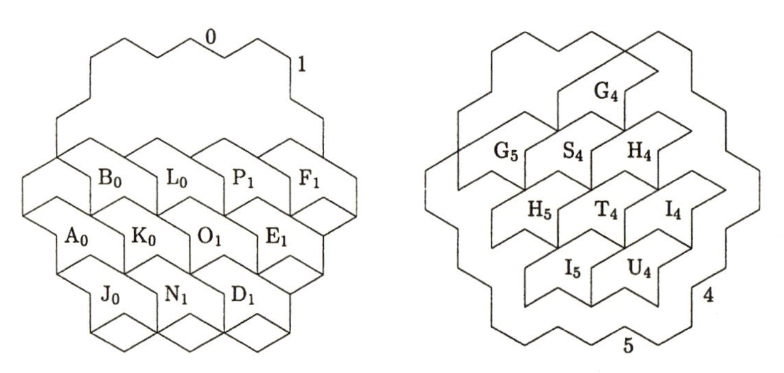

Figure 11. Continued: Coding the Crown's position.

direction its tail is pointing. Positions *s, t,* and *u* are symmetrical and carry only even subscripts; *c, g,* and *r* don't lead to legal solutions.

Now that you know how to classify solutions, note that O'Beirne's first solution (Figure 3) is of type $hE_5p_4J_5a_5$. Remember that the code doesn't specify the solution completely. There are at least ten solutions of class $hA_9p_4B_0f_0$, for example. (Can you find a larger class?)

How many solutions are there? A wild guess, based on how rarely duplicates appear, is about 50,000. There are already more than 4200 in the collection, which will be deposited in the Strens Collection in the Library at the University of Calgary. It isn't very easy to give a good upper bound. Coloring arguments don't seem to lead to much restriction, but perhaps some reader will be more perspicacious.

There are 508 essentially different relative positions for the Chevron and Hexagon that have not been proved to be impossible, although some of these turned out to be so. We have found 247 of these cases (no fewer

Figure 12. Coding the position of the Lobster.

than twenty-six bit the dust during the writing of this article). With the Chevron in positions *g, h,* and *o,* there are respectively 21, 39, and 1 legal positions for the Hexagon, and solutions are known in all of these cases. As we go to press, Marc Paulhus has established that there are just

$$13 + 17 + 16 + 14 + 18 + 21 + 39 + 29 + 27 + 30 \\ + 28 + 30 + 1 + 23 + 33 + 7 + 33 + 22 + 32 = 433$$

different relative positions for the Chevron and Hexagon that yield solutions.

In 1966 Bert Buckley, then a graduate student at the University of Calgary, suggested looking for solutions on a machine. I didn't think he would be successful, but after a few months of intermittent CPU time on an IBM 1620 he found half a dozen or so solutions from which it was possible to deduce another fifty by hand. Figure 13 shows two of Bert Buckley's machine-made solutions.

With such a plethora of solutions, the discerning solver will soon wish to specialize. For example, how symmetrical a solution can you get? Figure 14 shows two solutions, the first found by John Conway and Mike Guy in

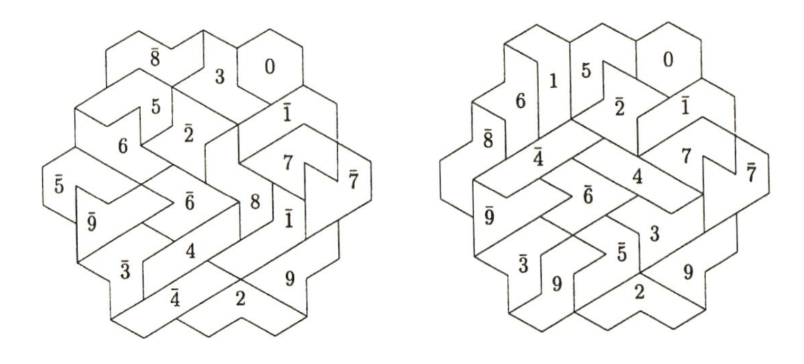

$hA_1n_xB_0o_3$ 66-05-30 $hA_1q_xF_9o_3$ 66-08-15

Figure 13. Solutions found by machine by Bert Buckley.

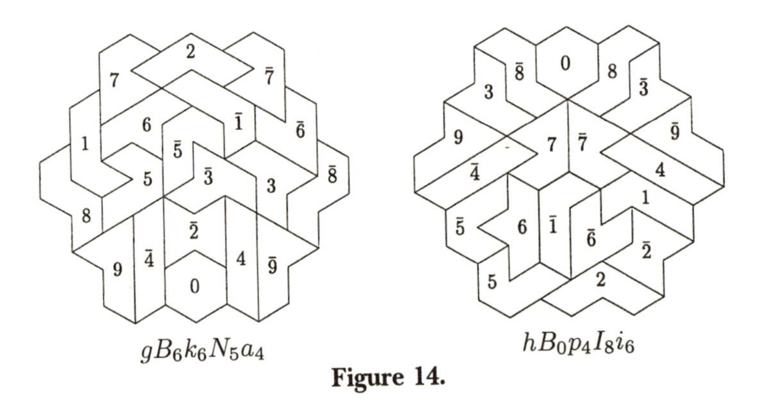

$gB_6k_6N_5a_4$ $hB_0p_4I_8i_6$

Figure 14.

1963, the second by the present writer, each with as many as 11 of the 19 pieces symmetrically placed.

Notice that there are two different kinds of axes of symmetry. All edges of the pieces lie in one of three different directions, at angles of 60 degrees to one another. We have always drawn the hexiamond with one of these directions vertical, but you may prefer to put one of them horizontal; Figure 15 shows eight solutions, each with 11 symmetrically placed pieces, found by John Conway in 1964. To deduce the others from the single diagram shown, rearrange pieces $2\bar{2}9\bar{9}$ or $3\bar{3}8\bar{8}$ and note that the hexagon formed by $01\bar{1}6\bar{6}$ has the symmetries of a rectangle.

Conway showed that you can't have more than 15, respectively 13, pieces placed symmetrically with respect to the two kinds of axis. It seems unlikely that anyone can beat 11.

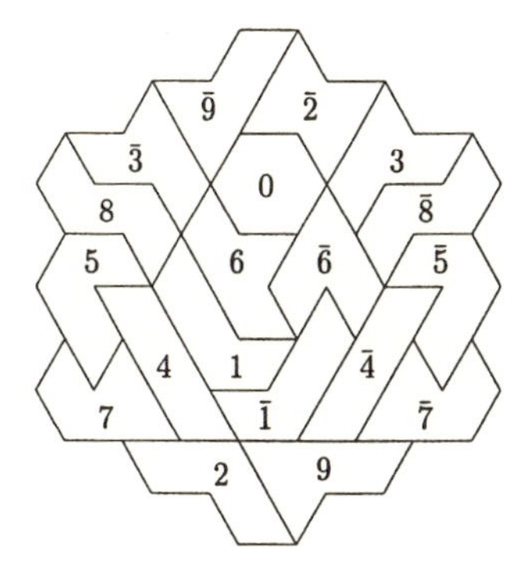

Figure 15. $hE_\varepsilon i_0 O_0 l_\varepsilon$ The most symmetrical solution?

Some solutions come in large groups. For example, in Figure 14, rearrange pieces $\bar{2}3\bar{3}$, or $\bar{1}\bar{3}5\bar{5}6$, after which you can swap $0\bar{2}$ with $\bar{3}5$. You can rotate $0\bar{2}\bar{4}$ ($\bar{3}\bar{4}\bar{5}$) or $0\bar{2}4$ ($\bar{3}4\bar{5}$) and produce two "keystones," which can be swapped, and so on.... Figure 16 shows a solution displaying three keystones that can be permuted to give other solutions. Figure 17, found by Mike Guy, shows that keystones ($4\bar{4}\bar{9}$) can occur internally and don't have to fit into a corner.

Figure 18 has the Hexagon in position *G* and the other pieces forming three congruent sets. Figure 19 is one of at least eight examples that have six pieces meeting at a point, here $\bar{3}5\bar{5}\bar{6}7\bar{7}$. Readers will no doubt discover other curiosities and find their own favorites.

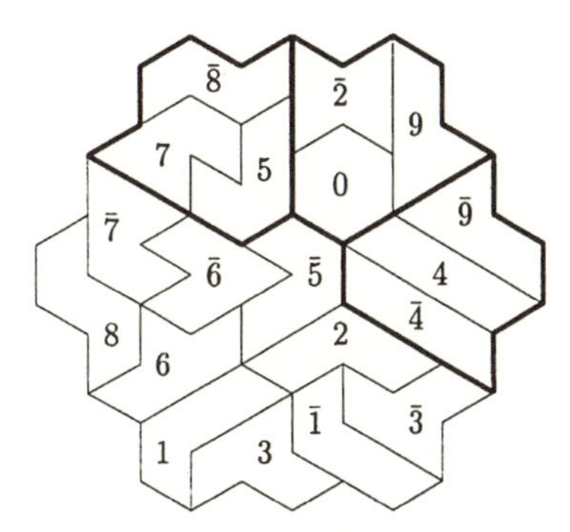

Figure 16. $jE_1 p_0 B_x s_2$ has three keystones.

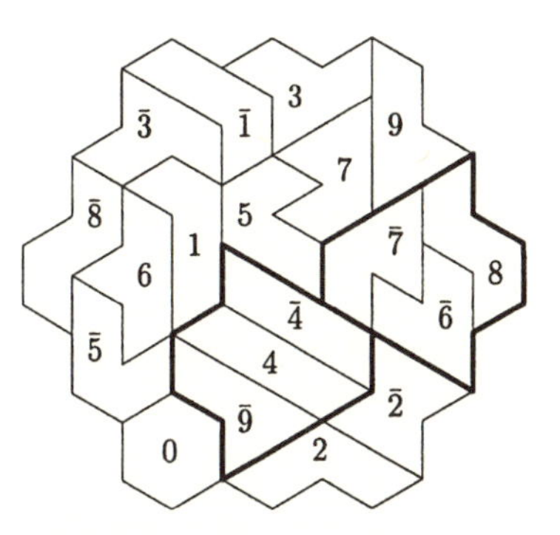

Figure 17. $hA_7 p_4 U_8 f_4$ has an internal keystone.

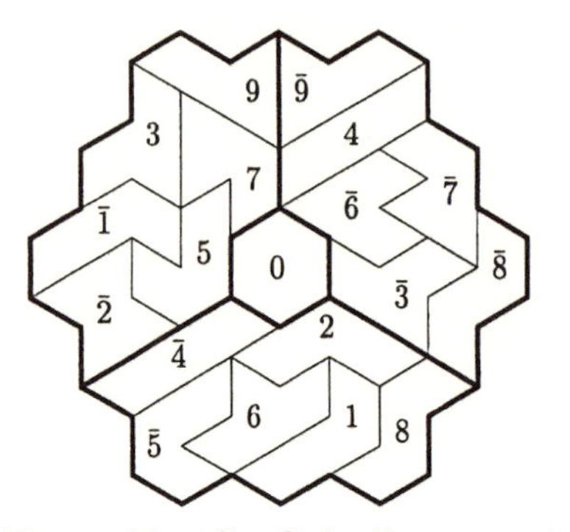

Figure 18. $jGp_8G_xk_9$ (3-symmetry).

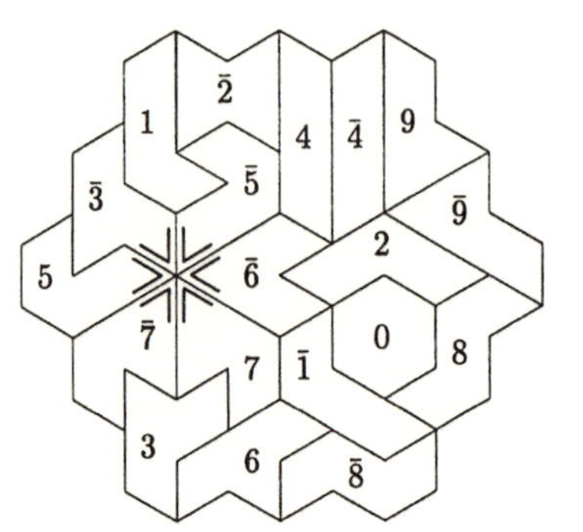

Figure 19. $qD_4i_0F_5t_8$
Six pieces meet at a point.

In an earlier draft I wrote that with modern search techniques and computing equipment and some man–machine interaction it has probably become feasible to find all the solutions of the Hexiamond. In May 1996 Marc Paulhus wrote a program that used only a few days of computer time to find all the solutions.

Their numbers, classified according to the position of the Chevron, are

b	c	d	e	f	g	h	i	j	k
130	195	533	193	377	2214	111,460	584	985	885

l	m	o	p	q	r	s	t	u	Total
637	914	749	498	1238	31	1537	264	1094	124,518

and, by the Hexagon's position

A 75,489 B 15,717 C 6675 D 7549 E 11,447 F 5727 G 1914

I don't think that this need take the fun out of one of the best two-dimensional puzzles ever invented. On the contrary, for those who prefer their puzzles to have just one answer, there are no fewer than 40 relative positions of the Chevron and Hexagon that determine such a unique solution:

bC_1	bC_3	bD_0	bE_3	cB_2	cC_1	cC_2	cC_5	cC_6	cD_6
cF_0	eC_3	eC_5	fC_6	fF_2	iC_1	iC_x	jD_0	kC_8	kC_9
lD_x	lF_6	mC_5	pF_8	pF_x	pG	qC_0	qF_x	rB_8	sC_x
sC_ε	tB_8	tB_x	tC_1	tC_x	tE_1	tE_7	tF_0	tF_6	tF_x

How long will it take you to find them all without peeking at Marc's database [5]?

After writing a first draft of this article, I wrote to Kate Jones, president of Kadon Enterprises and discovered that she markets a set of "iamonds" under the trademark "Iamond Ring." For this purpose the iamonds are not counted as different if they are reflections of one another, so, if you want to use them to make O'Beirne's hexiamond puzzle, you must get two sets. The numbers of iamonds, diamonds, triamonds, ... , are given in the following table, where the last line counts reflections as different:

Order	1	2	3	4	5	6	7
# of iamonds	1	1	1	3	4	12	24
with reflections	1	1	1	4	6	19	44

The Iamond Ring is a beautiful arrangement, discovered by Kate herself. Figure 20, in which each *x*-iamond is labeled $x, 1 \leq x \leq 7$, doesn't do justice to the very appealing colored pieces of the professionally produced puzzle.

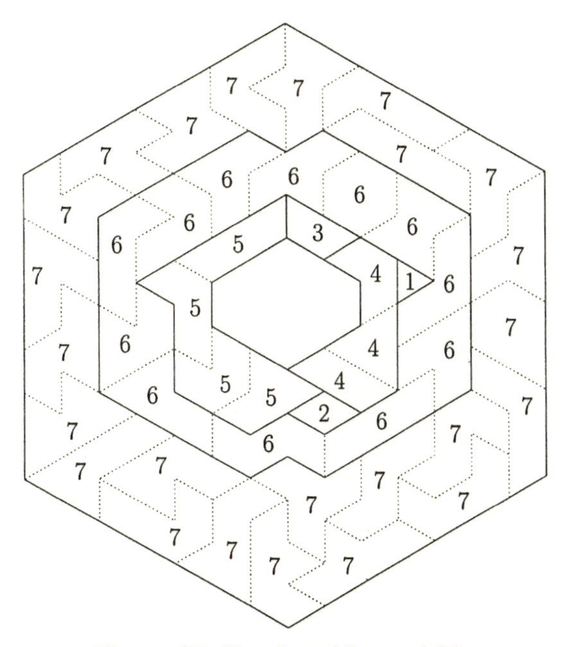

Figure 20. Kate Jones' Iamond Ring.

References

[1] E. R. Berlekamp, J. H. Conway, and R. K. Guy, *Winning Ways for Your Mathematical Plays,* Academic Press, London, 1982, pp. 787–788.

[2] Richard K. Guy, Some mathematical recreations I, *Nabla (Bull. Malayan Math. Soc.),* 9 (1960) pp. 97–106; II pp. 144–153; especially pp. 104–106 and 152–153.

[3] George E. Martin, *Polyominoes: A Guide to Puzzles and Problems in Tiling,* Math. Assoc. of America Spectrum Series, 1991, pp. 168–170.

[4] Thomas H. O'Beirne, *Puzzles and Paradoxes,* Oxford University Press, 1965; see Puzzles and Paradoxes No. 44: Pentominoes and hexiamonds, *New Scientist,* 259 (61-11-02) pp. 316–317.

[5] Marc Paulhus, *A database for O'Beirne's Hexiamond,* submitted.

Japanese Tangram: The Sei Shonagon Pieces

Shigeo Takagi

In 1974, I received a letter from Kobon Fujimura, a famous puzzlist in Japan. I heard that Martin Gardner had been planning to write about tangrams, so I sent a report about the Japanese tangram to him.

The tangram came to Japan from China in the early 19th century, and the Japanese edition of *Qiqiaotu Hebi (The Collected Volume of Patterns of Seven-Piece Puzzles)* (1813) was published in 1839.

In fact, Japan had a similar puzzle already. In 1742, a little book about Japanese seven-piece puzzles was published. The book was called *Sei Shonagon Chie-no-ita (The ingenious pieces of Sei Shonagon).* Sei Shonagon, a court lady of the late 10th and early 11th centuries, was one of the most clever women in Japan. She was the author of a book entitled *Makura no Soshi (Pillow Book: A Collection of Essays).*

(a) Tangram pieces. (b) The Sei Shonagon pieces.

Sei Shonagon Chie-no-ita is a 32-page book, 16 centimeters wide and 11 centimeters long. The introduction bears the pseudonym Ganreiken, but nobody knows its real author. There are 42 patterns with answers, but their shapes are inaccurate. The puzzle was introduced to the world by Martin Gardner in his book *Time Travel*[1], and he elaborated as follows.

"Shigeo Takagi, a Tokyo magician, was kind enough to send me a photocopy of this rare book. Unlike the Chinese tans, the Shonagon pieces will form a square in two different ways. Can you find the second pattern? The pieces also will make a square with a central square hole in the same orientation. With the Chinese tans it is not possible to put a square hole anywhere inside a large square."

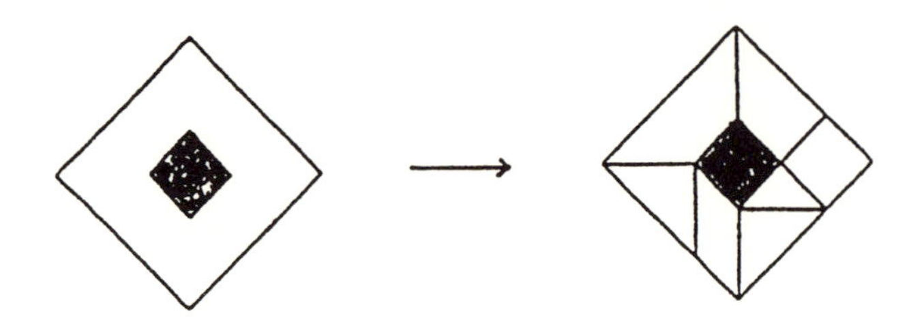

A square with a center hole

I know of two other books that are collections of patterns of the Sei Shonagon pieces. In about 1780, Takahiro Nakada wrote a manuscript entitled "Narabemono 110 (110 Patterns of an Arrangement Pattern)," and *Edo Chie-kata (Ingenious Patterns in Edo)* was published in 1837. In addition, I possess a sheet with wood-block printing on which we can see patterns of the Sei Shonagon pieces, but its author and date of publication are unknown.

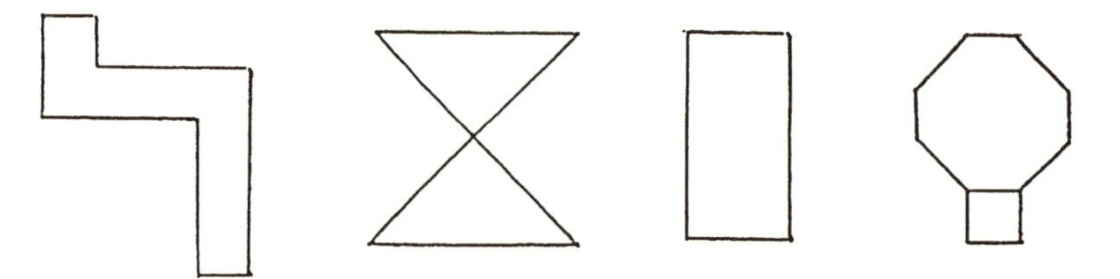

Some patterns formed from the Sei Shonagon pieces

[1] 1988, W.H. Freeman and Co., NY

How a Tangram Cat Happily Turns into the Pink Panther

Bernhard Wiezorke

Do you want to create, or better generate, a new two-dimensional puzzle? Nothing could be easier! Just take one of the many well-known puzzles of this type and submit it to a geometrical transformation. The result will be a new puzzle, and – if you do it the right way – a nice one. As an example, let us take the good old Tangram and try a very simple transformation, a linear extension (Figure 1).

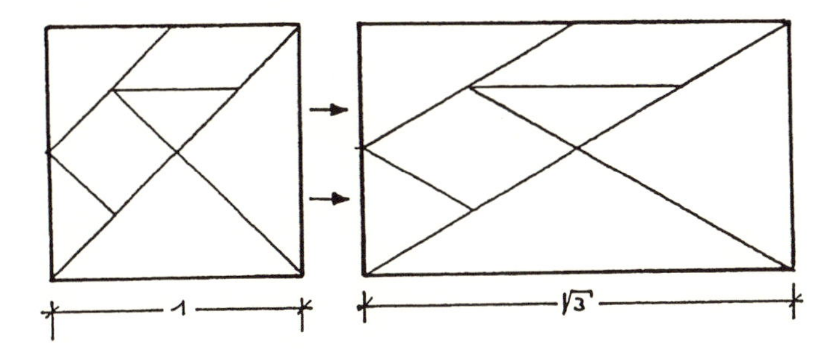

Figure 1. Tangram submitted to a linear extension. The result is a set of pieces for a new puzzle, a Tangram derivative called Trigo-Tangram.

While the tangram pieces can be arranged in an orthogonal grid, the pieces of our new puzzle, due to $\sqrt{3}$ as extension factor, fit into a grid of equilateral triangles, a trigonal grid. That's why I gave it the name Trigo-Tangram.

Puzzles generated by geometrical transformations I call *derivatives*; our new puzzle is a Tangram derivative, one of an infinite number. Depending on the complexity of the transformation, certain properties of the original puzzle are preserved in the derivative, others not. Tangram properties preserved in Trigo-Tangram are the linearity of the sides, the convexity of the pieces, and the side length and area ratios. Not preserved, for example,

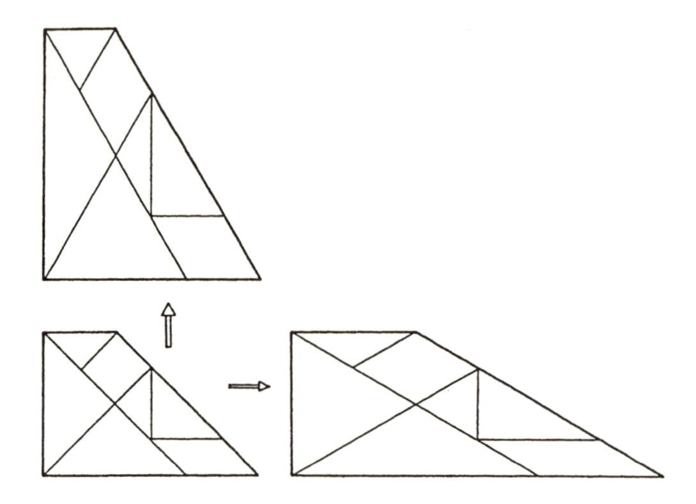

Figure 2. Suitable problems for a puzzle derivative can be found by transforming problems of the original puzzle. For Trigo-Tangram, two problems can be generated by extending this original Tangram problem in two different directions. The transformation of the solution can be a solution for the derivative (right image) or not (upper image).

are the angles and the congruence of both the two big and the two small triangles.

For our derivatives, we do not only need the pieces, we need problems as well. Problems? No problem at all! We just take some figures for the original puzzle and submit them to the same transformation as the puzzle itself. The images of the figures will be suitable problems for the derivative. In Figure 2 this is done for a Tangram problem and its solution. As you can see, the corresponding linear extension can be performed in two different directions, which gives us two different problems. And you can see another important fact: The transformation of the solution does not always render a solution for the derivative.

Figure 3 shows some problems for Trigo-Tangram. So, take a piece of cardboard, copy and cut out the pieces, and enjoy this new puzzle.

Here is another way to have fun with a puzzle and its derivative: Take the original Tangram, lay out a figure, and try to make a corresponding figure with the pieces of Trigo-Tangram. In many cases, you get funny results. Figure 4 shows how a Tangram bird stretches its neck and how a Tangram cat turns into the Pink Panther.

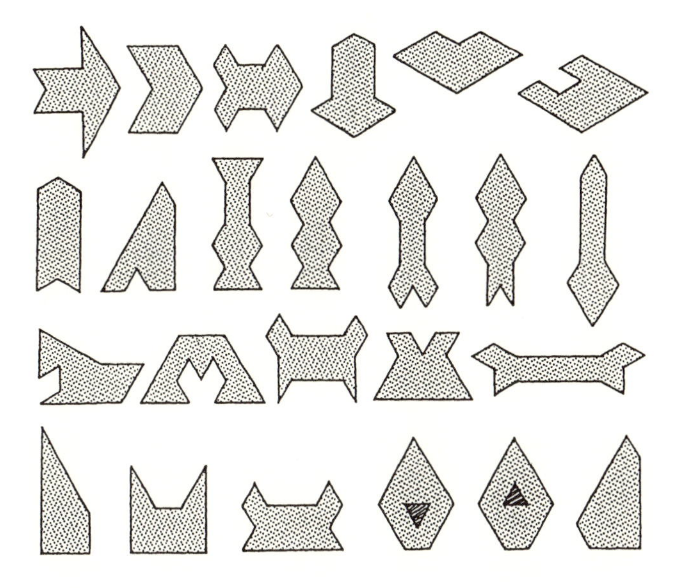

Figure 3. Some problems for the Trigo-Tangram puzzle.

Transforming two-dimensional puzzles provides endless fascination. Just look what happens if you rotate the Tangram in Figure 1 by 90° and then perform the transformation. The square and parallelogram of the original are transformed into congruent rhombi, which makes the derivative less convenient for a puzzle. You obtain better results by applying linear extensions diagonally (Figure 5). Since in this case the congruence between the two large and two small Tangram triangles is preserved, the two puzzles generated are even more Tangram-like than the Trigo-Tangram. To create

Figure 4. A Tangram bird stretches its neck, and a Tangram cat turns into the Pink Panther.

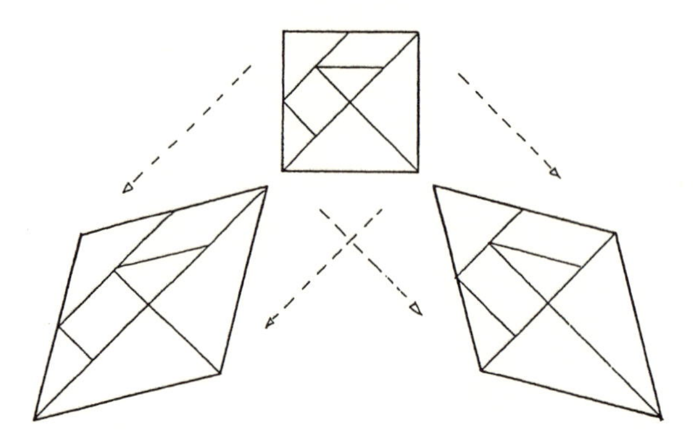

Figure 5. Transforming the Tangram by diagonal extensions leads to two derivatives which are more Tangram-like than the one shown in Figure 1.

problems for these derivatives is left to the reader. One set of problems will serve for both of them.

I hope I have stimulated your imagination about what can be done by transforming puzzles so you may start your own work. Take your favorite puzzle, and have fun generating individual derivatives for yourself and your friends!

Polly's Flagstones

Stewart Coffin

I wish to report on some recent correspondence with my good friends the Gahns in Calcutta. You may remember meeting Paul in *The Puzzling World of Polyhedral Dissections.* His wife Polly is an avid gardener. She presented me with the following problem.

Polly places precisely fitted flagstones around various plantings in her garden in order to suppress weeds. She and Paul have a bent for geometrical recreations, so they are always on the lookout for creative and original solutions to their various landscaping projects. She had one large stone that was perfectly square. She asked me if it were possible to cut the stone into four pieces that could be arranged to form a somewhat larger square perimeter enclosing a square hole having sides one quarter those of the original stone. The simple solution to this problem which I then sent to her, a classic dissection of the square, is shown below.

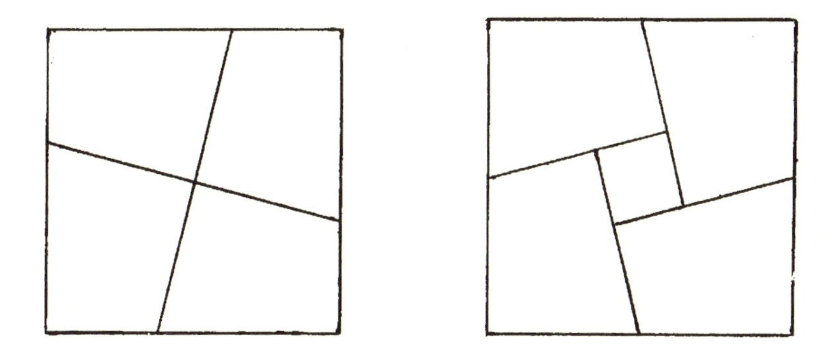

In order to meet Polly's required dimensions, the sides are divided in the ratio of 3 to 5. Frankly, I was surprised that Polly would bother to write about such a trivial dissections problem, but I should have known better. She wrote back with profuse thanks for such an "elegant" solution, and her sarcasm rather put me on guard.

Now she posed another problem. She decided that she preferred mostly rectangular rather than square planting spaces. Suppose we were to consider starting with a large rectangular stone, which she proposed that we cut into four pieces, but this time to form a rectangular perimeter enclosing a rectangular planting space. And for good measure, to allow for more flexibility in landscaping, how about a scheme whereby the stones could be rearranged to form any one of three different-sized rectangular openings, each one enclosed by a rectangular perimeter. This required a bit more reflection than the first problem, but after tinkering for a while with paper, pencil, and scissors, I came up with the scheme shown below.

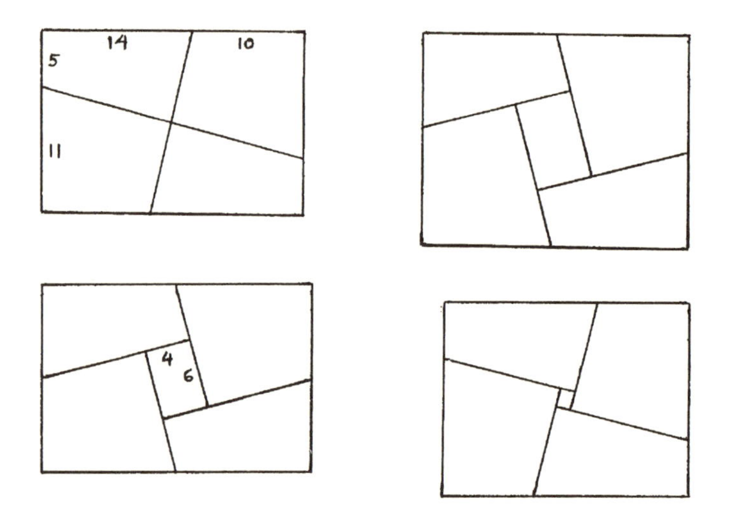

Almost any rectangle dissected symmetrically by two mutually perpendicular lines that touch all four sides will produce four quadrilaterals that can be rearranged to create the required three different rectangular holes enclosed by rectangles. One of these rectangular holes will have the same shape as the original rectangle and will be surrounded by a rectangle also having that same shape. With the dimensions shown, the medium-sized rectangular hole will have dimensions exactly one quarter those of the original stone. The smaller and larger holes have dimensions that are irrational.

If you think that Polly was satisfied with this solution, then you don't know Polly very well. She immediately wrote back and suggested that it was a shame to cut up two beautiful large stones when one might do, cut into four pieces, which could then be rearranged to create a square hole within a square enclosure or any one of three different rectangular holes within rectangular enclosures. It was then I realized that from the start she had just been setting me up. I decided to play into it, so I wrote back and asked what made her so sure it was even possible. Sure enough, by return mail arrived her solution, shown below.

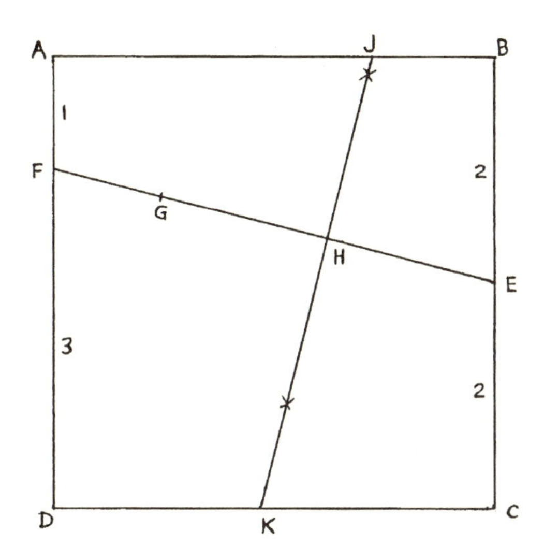

For the purpose of this example, let the original square stone *ABCD* be four feet on a side. Locate the midpoint *E* of side *BC*. The location of point *F* is arbitrary, but for this example let it be one foot from *A*. Draw *EF*. Subtract length *CE* from length *DF*, and use that distance from *F* to locate point *G*. Bisect *EG* to locate *H*, and draw *JHK* perpendicular to *EHF* (most easily done by swinging arcs from *E* and *G*).

Don't ask my why this works, but it does. The actual arrangements of the pieces for the four different solutions, one square and three rectangular, are left for the reader to discover. One of the rectangles has a pair of solutions – the others are unique. With these dimensions, the square hole will be one foot square. For added recreation, note that the pieces can also be arranged to form a solid parallelogram (two solutions), a solid trapezoid (two solutions), and a different trapezoid (one solution).

An interesting variation is to let points *A* and *F* coincide, making one of the pieces triangular. The same solutions are possible, except that one of the trapezoids becomes a triangle.

Now what sort of scheme do you suppose Polly will come up with for that other stone, the rectangular one?

Those Peripatetic Pentominoes

Kate Jones

This recital is yet another case history of how the work of one man—Martin Gardner—has changed the course and pattern of a life.

In 1956 I received a gift subscription to *Scientific American* as an award for excellence in high school sciences. My favorite part of the magazine was the "Mathematical Games" column by Martin Gardner. The subscription expired after one year, and I was not able to renew it. It expired one month before the May 1957 issue, so I did not get to see the historic column introducing pentominoes (Figure 1) to a world audience.

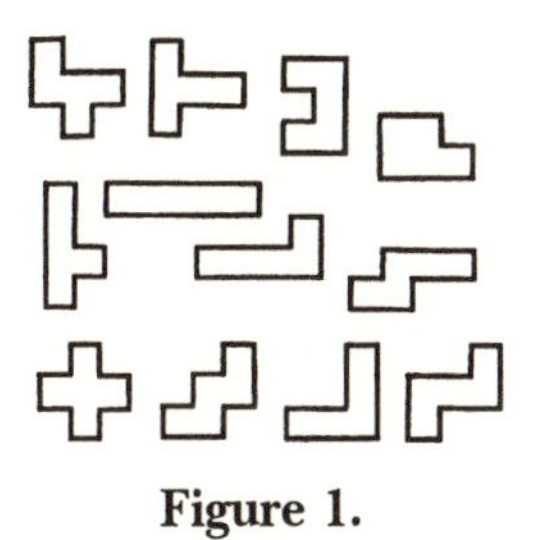

Figure 1.

That column was inspired by a 1954 article in the *American Mathematical Monthly*, based on Solomon Golomb's presentation in 1953 to the Harvard Mathematics Club. Golomb's naming of the "polyomino" family of shapes and their popularization through Martin Gardner's beloved column, focusing especially on the pentominoes, created an ever-widening ripple effect.

Arthur C. Clarke, in his mostly autobiographical *Ascent to Orbit*, declares himself a "pentomino addict", crediting Martin Gardner's column as the source. Thwarted from including pentominoes in the movie *2001: A Space Odyssey*, as the game HAL and Bowman play (the film shows them playing chess), Clarke wrote pentominoes into his next science fiction book, *Imperial Earth*. Later editions of the book actually show a pentomino rectangle on the flyleaf.

In late 1976, a group of expatriates stationed in Iran took a weekend trip to Dubai. As we loitered around the airport newsstand, a paperback rack with a copy of *Imperial Earth* caught my eye. A longtime Clarke fan, I bought the book and soon caught pentomino fever. It was a thrill to find mention of Martin Gardner in the back of the book.

Playing first with paper and then with cardboard was not enough. I commissioned a local craftsman to make a set of pentominoes in inlaid ivory, a specialty of the city of Shiraz. This magnificent set invited play, and soon friends got involved. Exploration of the pentominoes' vast repertoire of tricks was a fine way to spend expatriate time, and inevitably a domino-type game idea presented itself to me, to be shared with friends.

Fast forward to December 1978, when the Iranian revolution precipitated the rapid evacuation of Americans. Back home I was in limbo, having sold my graphics business and with no career plans for the future. My husband's job took care of our needs, but idleness was not my style. A friend's suggestion that we "make and sell" that game I had invented popped up just then, and after some preliminary doubts we went for it.

By fall of 1979 a wooden prototype was in hand, and the pieces turned out to be thick enough to be "solid" pentominoes. Well, of course that

Figure 2. Super Quintillions.

was the way to go! A little research turned up the fact that "pentominoes" was a registered trademark of Solomon Golomb, so we'd have to think of another name. Fives ... quints ... *quintillions!* It was with great pride, joy and reverence that we sent one of the first sets to the inspiration of our enterprise, Martin Gardner. And it was encouraging to us neophyte entrepreneurs when *Games* magazine reviewed Quintillions and included it on the "Games 100" list in 1980.

We had much to learn about marketing. The most important lesson was that one needed a product "line", not just a single product. And so Quintillions begat a large number of kindred puzzle sets, and most of them sneak in some form of pentomino or polyomino entity among their other activities. Here (chronologically) are the many guises and offspring of the dozen shapes that entered the culture through the doorway Martin Gardner opened in 1957.[1]

Super Quintillions: 17 non-planar pentacubes (plus one duplicate piece to help fill the box). These alone or combined with the 12 Quintillions blocks can form double and triplicate models of some or all of the 29 pentacube shapes (see Figure 2).

Leap: a 6×6 grid whereupon polyomino shapes are plotted with black and white checkers pieces in a double-size, checkerboarded format. The puzzle challenge: Change any one into another in the minimum number of chess knight's moves, keeping the checkerboard pattern (Figure 3).

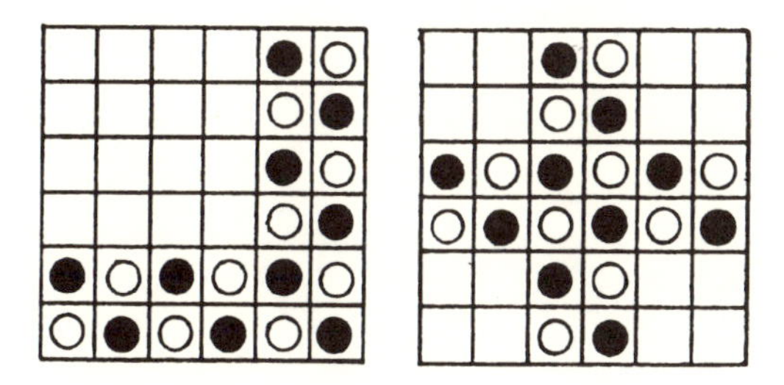

Figure 3.

[1]All these are products made by Kadon Enterprises, Inc., and the product names are trademarks of Kadon.

Void: a 4×4 grid on which a "switching of the knights" puzzle is applied to pairs of polyomino shapes from domino to heptomino in size, formed with checkers (Figure 4). What is the minimum number of moves to exchange the congruent groups of black and white knights?

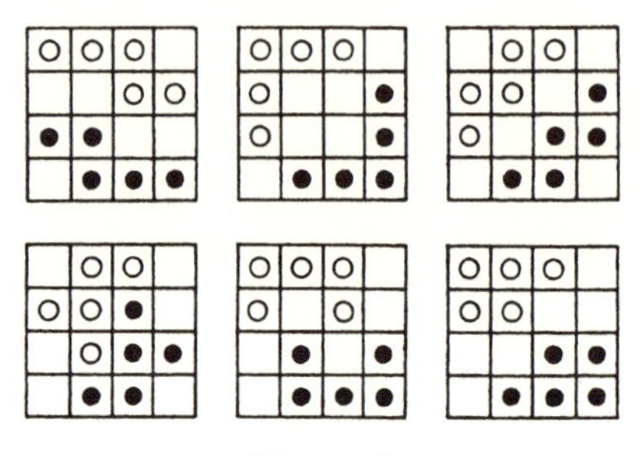

Figure 4.

Quintachex: the pentominoes plus a 2×2 square checkerboarded on both sides (different on the two sides). The pieces can form duplicate and triplicate models of themselves, with checkerboarded arrangements (Figure 5).

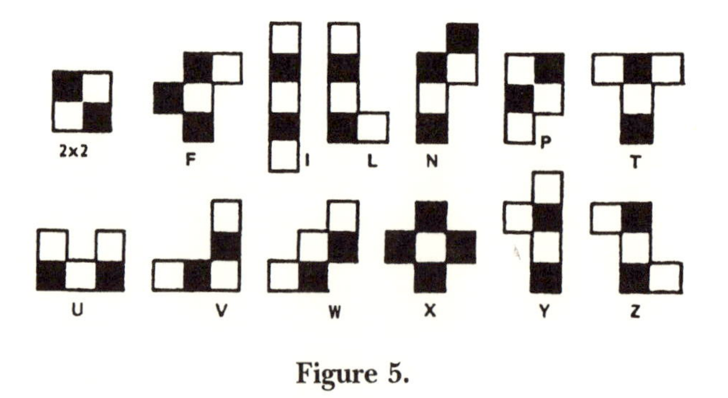

Figure 5.

Colormaze: Square tiles in six colors form double polyomino shapes with no duplicate color in any row, column, or diagonal. In another use, double polyominoes can be formed with 2×2 quadrants of each color and then dispersed through a maze-like sequence of moves to the desired final position (Figure 6).

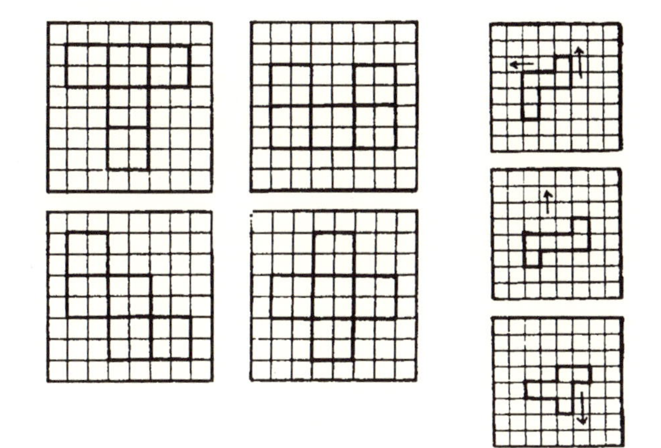

Figure 6.

Poly-5: two-dimensional polyominoes of orders 1 to 5, in a four-way symmetrical tray of 89 unit squares (Figure 7).

Figure 7.

Sextillions: the hexomino shapes can form double through sextuple copies of themselves and various enlargements of the smaller polyominoes. Size-compatible with Poly-5.

Snowflake Super Square: the 24 tiles are all the permutations of three contours–straight, convex, concave–on the four sides of a square. They can form various single- and double-size polyomino shapes (Figure 8).

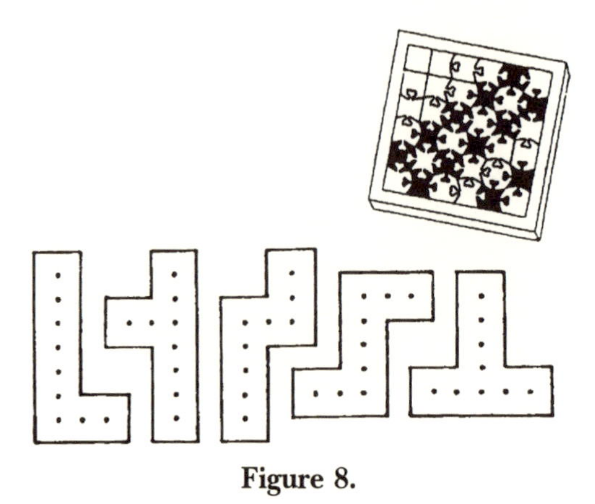

Figure 8.

Triangoes: orders 1 and 2 polytans permuted with two or more colors (square, triangle, parallelogram). The pieces can form diagonally doubled pentominoes and hexominoes with colormatching adjacency of tiles (Figure 9).

Figure 9.

Lemma: A matrix of multiple grids lends itself to polyomino packing with checkers in three colors.

Multimatch I: The classic MacMahon Three-Color Squares have been discovered to form color patches of pentominoes and larger and smaller polyominoes within the 4×6 rectangle and 5×5 square, including partitions into multiple shapes (Figure 10).

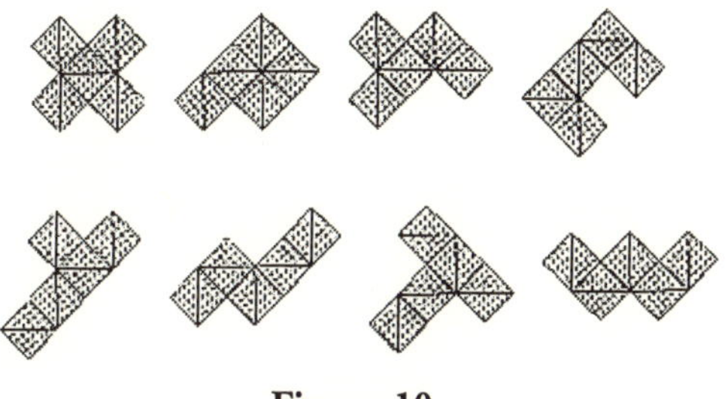

Figure 10.

Multimatch II: The 24 tri-color squares with vertex coloring (each tile is a 2×2 of smaller squares) can form polyominoes both as shapes with colormatching and as color patches on top of the tiles (Figure 11).

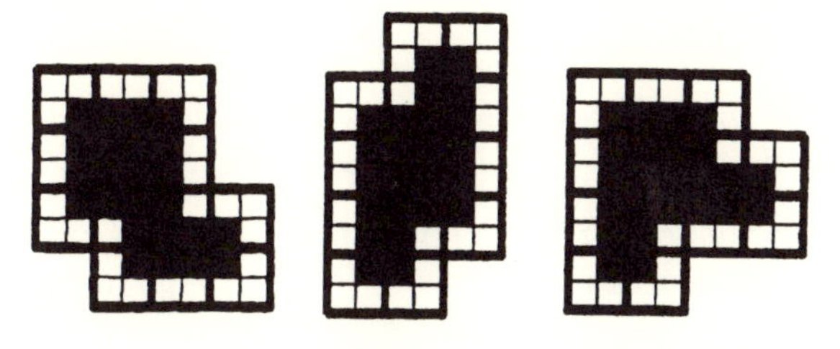

Figure 11.

Gallop: On a 6×12 grid, double checkerboard hexomino shapes defined by pawns are transformed with knight's moves (borrowed from the **Leap** set). Another puzzle is to change a simple hexomino made of six pawns into as many different other hexominoes as possible by moving the pawns with chinese-checker jumps (Figure 12).

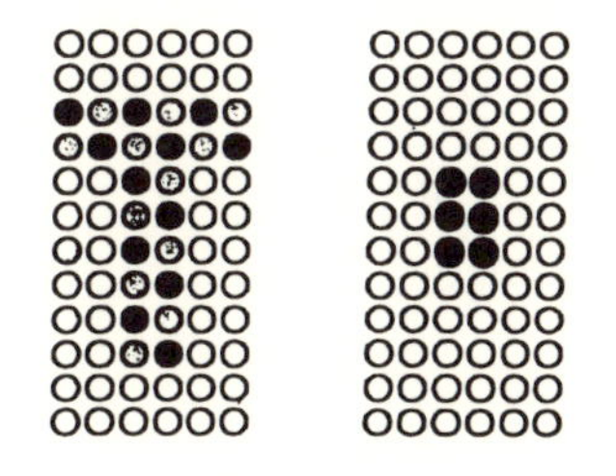

Figure 12.

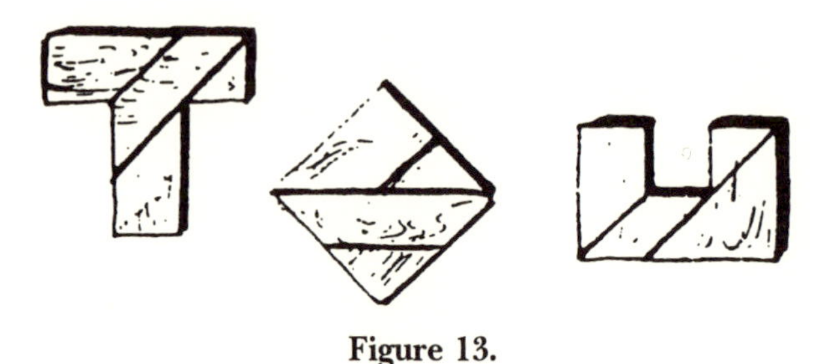

Figure 13.

Tiny Tans: Four triangle-based pieces can make some pentomino shapes. The T and U puzzles are actually dissected pentominoes (Figure 13).

Throw a Fit: Mulit-colored cubes form pairs of pentominoes with color-matching.

Perplexing Pyramid: A Len Gordon invention, the six pieces comprising 20 balls can make most of the pentominoes in double size.

Quantum: Pentomino and larger polyomino shapes are created by movement of pawns (the puzzles to be published as a supplement to the existing game rules).

Rhombiominoes: A skewed embodiment of pentominoes, where each square is a rhombus the size of two joined equilateral triangles. The 20 distinct pieces form a 10×10 rhombus (Figure 14). This is a limited-edition set.

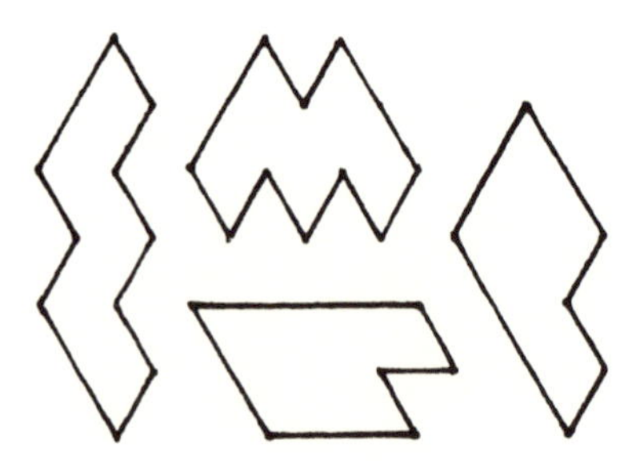

Figure 14.

Heptominoes: These sets of the 108 seven-celled polyominoes are produced by popular demand, sized to Sextillions and Poly-5.

Octominoes: The 369 eight-cell shapes exist in limited-edition sets, sized to smaller members of the family.

The Hexacube: The 166 planar and non-planar hexacube pieces plus 4 unit cubes form a $10 \times 10 \times 10$ cube. A vast territory just begging for exploration, this set is sized to Quintillions.

The above concepts arose and became possible because Martin Gardner fostered a love of mathematical recreations among his readers. A natural side-effect of his raising the consciousness of the public about the joys of combinatorial sets has been the proliferation of other sets, besides polyominoes: polyhexes, polyiamonds, polytans. All these are finding homes in the Kadon menagerie.

Throughout the years of Kadon's evolution, the inspiring and nurturing presence of Martin Gardner has been there, in the reprints of his columns, through articles in various publications, with kind and encouraging words in correspondence, and always helpful information. We were honored that Martin selected us to produce his Game of Solomon and the Lewis Carroll Chess Wordgame, using Martin's game rules. The charm, humor and special themes of these two games set them apart from all others, and we are dedicated to their care and continuance.

In retrospect, then, my personal career and unusual niche in life came about directly as a result of the intellectual currents and eddies created by one mind—Martin Gardner's—whose flow I was only too happy to follow. I cannot imagine any other career in which I would have found greater satisfaction, fulfillment and never-ending challenge than in the creation of

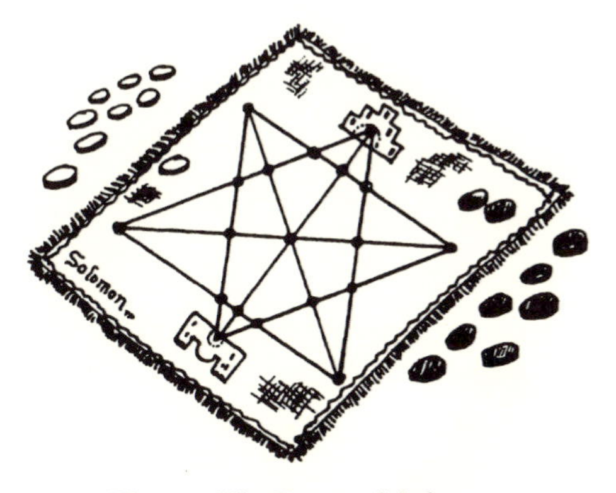

Figure 15. Game of Solomon.

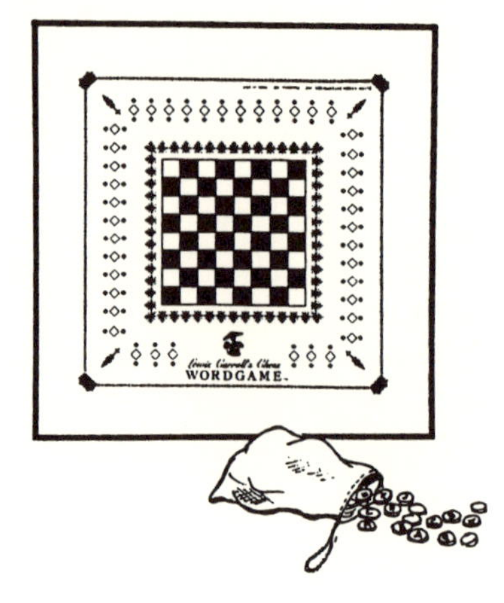

Figure 16. Lewis Carroll Chess Wordgame.

beautiful playthings for the mind. And assuredly, if there is a 90-degree angle or parallelogram to be found, the pentominoes will announce themselves in yet another manifestation. Thank you, Martin Gardner.

Self-Designing Tetraflexagons

Robert E. Neale

Flexagons are paper structures that can be manipulated to bring different surfaces into view. The four-sided ones discussed here are "flexed" by folding them in half along an axis, and then opening them up a different way to reveal a new face. Sometimes the opening is in a different direction (mountain or valley fold), sometimes along the other axis, and sometimes both. Although usually constructed from a strip of paper with attachment of the ends, flexagons seem more elegant when no glue or paste is required. Interesting designs of the faces can be found when the material is paper, or cardboard, with a different color on each side (as in standard origami paper). I call these "self-designing." What follows is the result of my explorations.

Although the text refers only to a square, the base for construction can be any rectangle. There are five different starting bases (paper shapes; see Figures 1–5). There are three different ways of folding the bases. Warning: Not all the results are interesting. But some are, my favorites being three: the first version of the cross plus-slit (for its puzzle difficulty); the first version of the square slash-slit (for the minimalist base and puzzle subtlety); and the third version of the square cross-slit (for its self-design). Note that the text assumes you are using two-colored paper, referred to, for simplicity, as black and white.

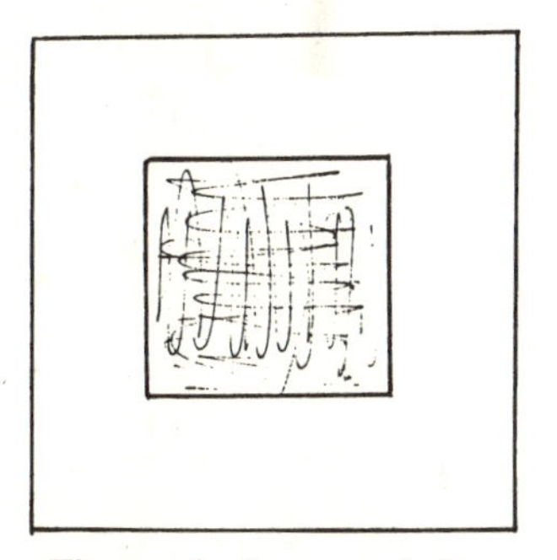

Figure 1. Square window.

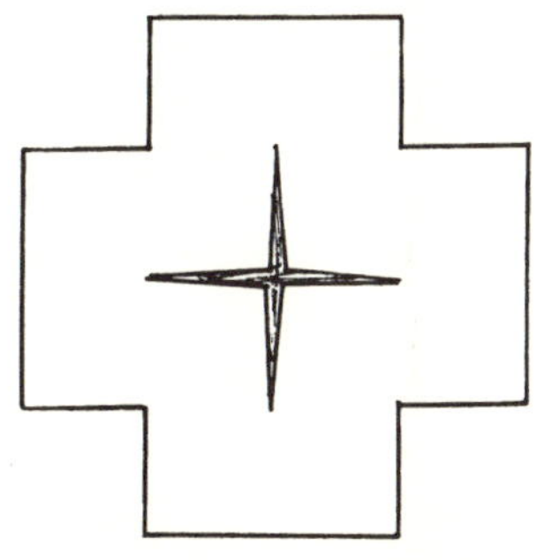

Figure 2. Cross plus-slit.

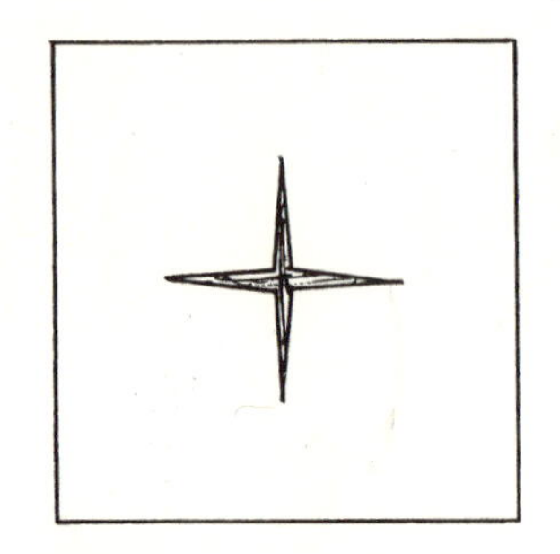

Figure 3. Square plus-slit.

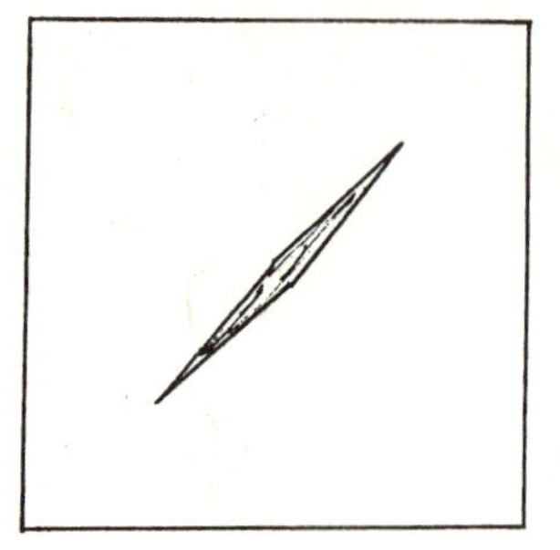

Figure 4. Square slash-slit.

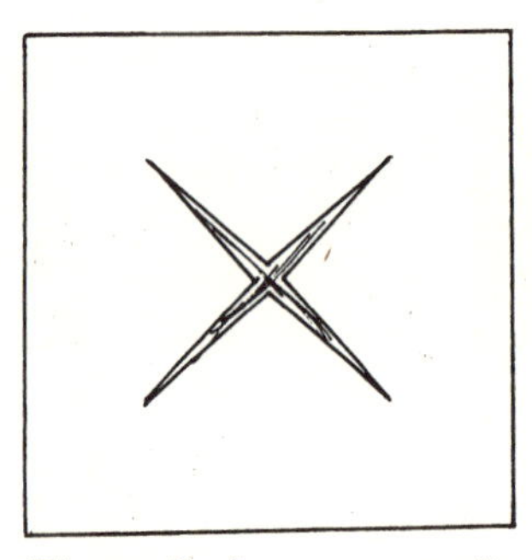

Figure 5. Square cross-slit.

Square Window

The first flexagon I saw that did not need glue was shown to me by Giuseppi Baggi years ago. It is a hexa-tetraflexagon made from a "window" – a square (of 16 squares) with the center (of 4 squares) removed (Figure 1). It has six faces that are easy to find.[1] I have recently used this window model for a routine about the riddle of the chicken and egg: "What's Missing Is What Comes First." This involves decorating the paper with markers, so is a separate manuscript.

It should be noted, however, that there are two ways of constructing this flexagon, both of which are discussed below.

Same Way Fold. As the dotted line indicates in Figure 6, valley fold the left edge to the center. The result is shown in Figure 7. As the dotted line indicates in Figure 7, valley fold the upper edge to the center. The result is

[1] Directions for constructing the hexa-tetraflexagon are found on pages 18–19 of Paul Jackson's *Flexagons,* B.O.S. Booklet No. 11, England, 1978.

shown in Figure 8. As the dotted line indicates in Figure 8, valley fold the right edge to the center. The result is shown in Figure 9.

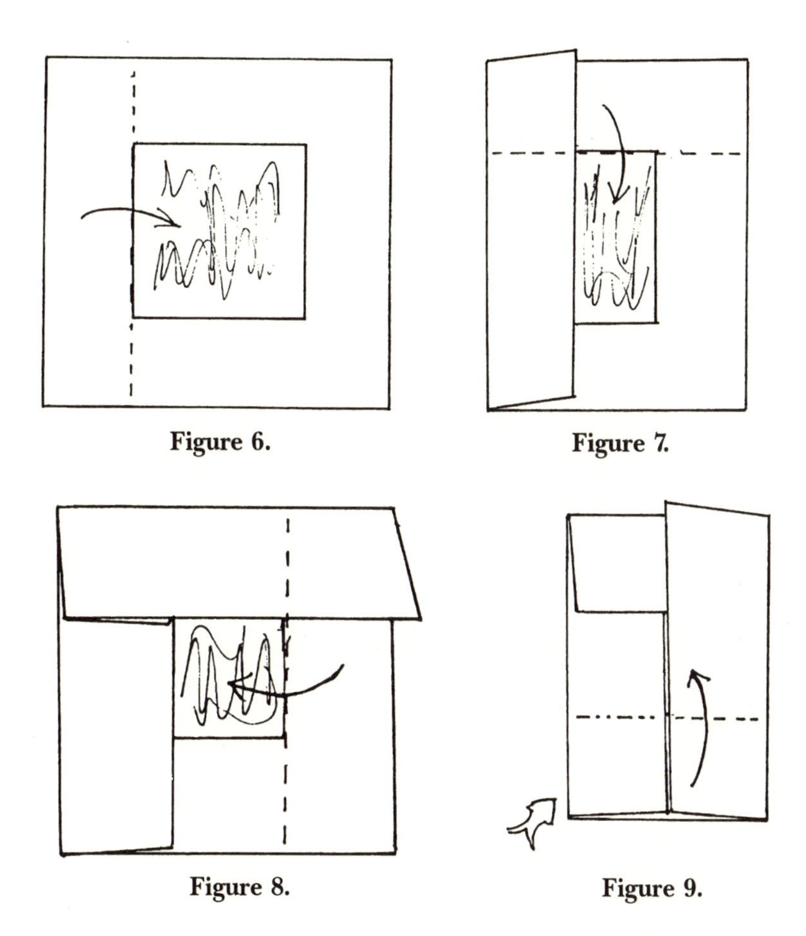

Figure 6.

Figure 7.

Figure 8.

Figure 9.

The final fold is very tricky to communicate, but not at all tricky to do when you understand it. The goal is to make this last corner exactly the same as the other three. As you would expect, the bottom edge will be valley folded to the center. The right half of this lower portion falls directly on top of the portion above it. The left half of this lower portion goes underneath the portion above it. (So Figure 9 shows a valley fold on the right half and a mountain fold on the left half.) To make this actually happen, lift up the upper layer only of the left half of the lower portion, and then fold the entire lower portion to the center. The result is shown in Figure 10.

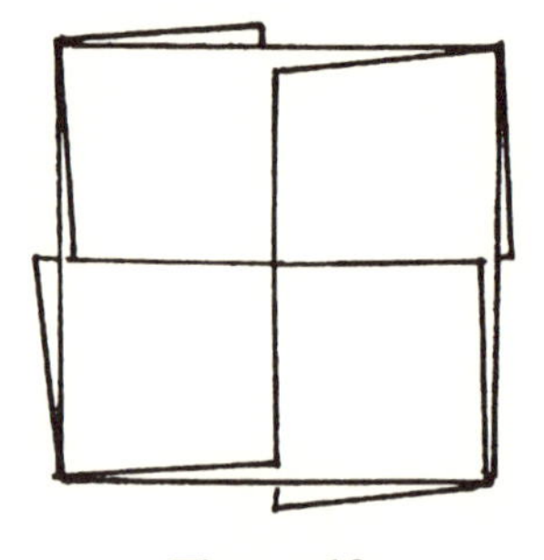

Figure 10.

Note that the model is entirely symmetrical. The four corners are identical, and both sides of the model are identical. (If this is not the case, you have made a mistake, probably on the last move.) This method amounts to folding the edges to the center, one after another (proceeding either clockwise or counterclockwise around the square), all of the folds being valley folds. This produces a Continuous Flex, moving in a straightforward manner from face 1 to 2 to 3 to 4 to 1, 5, 6, 1. The designs procured are merely three black faces and three white faces, as follows: black, black, white, white, black, black, white, white, black.

Alternating Way Fold. As Figure 11 indicates, mountain fold the left edge to the center. The result is shown in Figure 12. As the dotted line indicates in Figure 12, valley fold the upper edge to the center. The result is shown in Figure 13. As the dotted line indicates in Figure 13, mountain fold the right edge to the center. The result is shown in Figure 14.

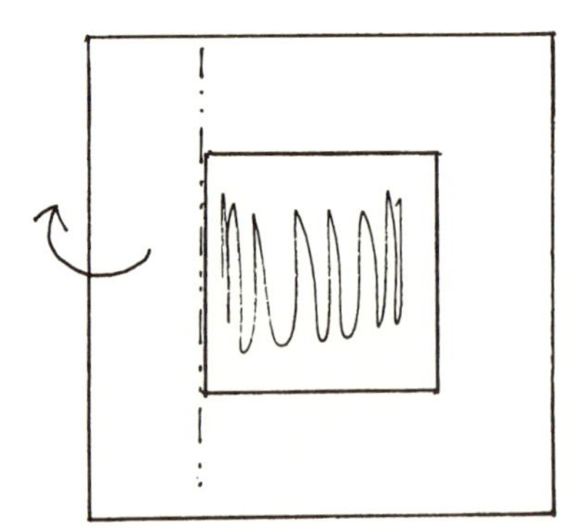

Figure 11.

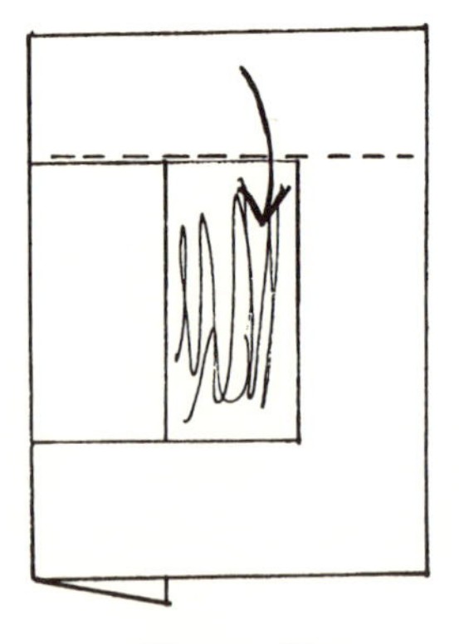

Figure 12.

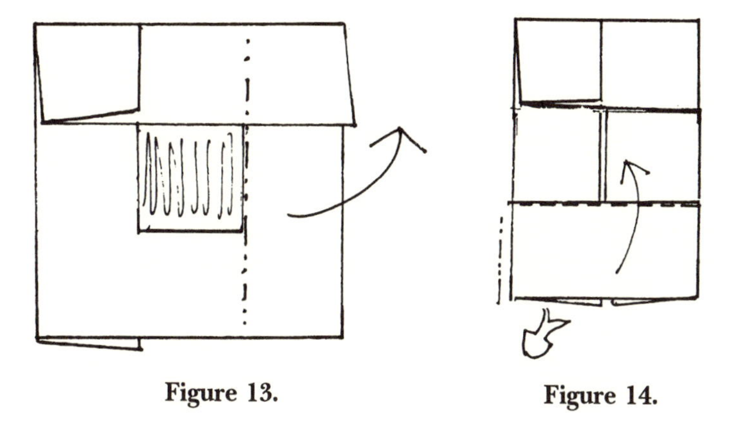

Figure 13. **Figure 14.**

Before you make the final valley fold of the bottom edge to the center, pull the bottom layer at the lower left corner away from the upper layer. Make the valley fold, and allow that bottom layer to go back to the bottom of the lower left corner. The result is shown in Figure 15. Note that opposite corners are identical, and both sides of the model are identical.

This noncontinuous Puzzle Flex can be flexed continuously through four faces only: 1 to 2 to 3 to 4 to 1, etc. Finding the other two faces is easy in this particular case, but backtracking is required. The four faces are identical: a checkerboard pattern of two black squares and two white squares. The other two faces are a solid black face and a solid white face.

Figure 15.

Cross Plus-Slit

The window model inspired me to think of other ways to form a flexagon without having to attach the ends to each other. One result was a puzzle made from a "cross" – a square (of 16 squares) with the four corner squares removed, and two slits in the center that intersect in the shape of a plus sign (see Figure 2).

This base is manipulated in a third way, the 3-D Tricky Way. This puzzle is interesting for two reasons: It is tricky to construct, and, while four of the faces are easy to find, the other two faces are quite difficult to discover, involving changing the flat flexagon into a three-dimensional ring and back again. The four faces are continuous and identical, being black and white checkerboard patterns. The other two faces are solid black and solid white. (Note: The trick fold for constructing the flexagon can be done in a slightly different way that has the four faces not continuous.)

3-D Tricky Fold In order to follow the instructions and understand the diagrams, number the base from one to six, exactly as indicated in Figure 16. Orient them just as indicated. The numbers in parentheses are on the back side of the base. You will make two, very quick, moves. The first renders the base 3-D, and the second flattens it again.

Note the two arrows in Figure 16. They indicate that you are to make the base 3-D by pushing two opposite corners down and away from you. So reach underneath and hold the two free corners of the 1 cells, one corner in each hand. When you pull them down and away from each other, they are turned over so you see the 5 cells. The other cells come together to form two boxes open at the top. There is a 6 cell at the bottom of each box

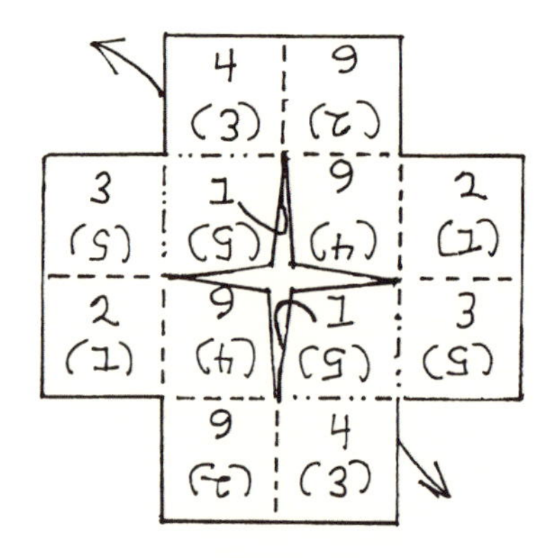

Figure 16.

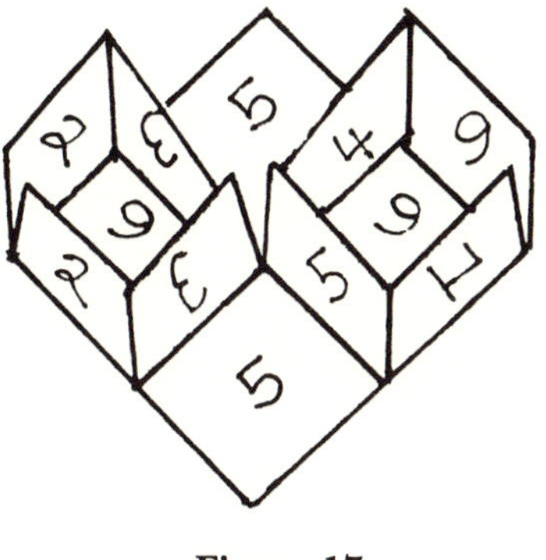

Figure 17.

(see Figure 17). Now the base should be flattened into a compact model with four cells showing on a side. Change your grip so you are holding the two inside corners of the 5 cells. (These are the corners opposite the ones you were holding.) Now push down on these corners, and at the same time, pull them away from each other. The 3-D boxes will collapse, the base forming a flat square of four cells. Once this happens, you should have four 1 cells facing you, and four 4 cells on the other side. The result is shown in Figure 18.

Sometimes, however, you will find that another number has appeared, a 2 instead of a 1, or a 3 instead of a 4. You can correct this easily by tucking the wrong cell out of sight, replacing it with the proper one. Check both sides. The model is completed.[2]

Flexing. You can find the faces numbered 1 to 4 by the usual flexing. Faces 5 and 6 are found by the following procedure. Begin with face 1 on the top and face 4 on the bottom. Mountain fold the model in half on the vertical axis, the left and right halves going back away from you. Do not open the model as you do when flexing. Rather, move the lower inside packet of squares (with 2 and 3 on the outside) to the left, and the upper inside packet of squares (with 2 and 3 on the outside also) to the right. Now open the model into a tube – a cube open at both ends. Collapse the tube in the opposite way. This creates a new arrangement. Flex it in the usual way to show face 5, then 3, and then make the tube move again

[2]Other directions for constructing this flexagon are on page 27 of Jackson's *Flexagons*, and in Martin Gardner's *Wheels, Life and Other Mathematical Amusements*, New York: W.H. Freeman & Co., 1983, pp. 64–68.

Figure 18.

to return to face 4 and then face 1. To find face 6, begin on face 4, with face 1 at the back, and repeat the moves just given.

Another version of this flexagon can be made by using the Same Way Fold. (Alternating Way gets you nowhere.) Four faces are easily found, but they are not continuous. Two faces can be found only by forming the rings, as mentioned above. The designs of the four faces are two solid black and two checkerboard. The other two faces are solid white.

Square Plus-Slit

About a year or so ago, I discovered a way to make a flexagon without removing any portion of the square. (The goal was inspired by my desire to make a flexagon from paper money with minimal damage.) It is made from a square with two slits intersecting the center in the shape of a plus sign (Figure 3).

For best results, use the Alternating Way fold. Flexing is continuous, and four faces can be shown: checkerboard, solid white, solid black, checkerboard. The Same Way Fold can be used also, but with poor results. Four faces can be shown, but they are not continuous. Indeed, both the third and fourth are derived from the first face. All faces are of the identical solid color. Using the 3-D Tricky Fold gives the same results as the Alternating Way Fold.

More recently, I have worked out some variations on the theme of removing no paper. These follow.

Square Slash-Slit

(Repeat "slash-slit" out loud quickly at your peril.) It is pleasing that a flexagon can be constructed from a square, or rectangle, with a single slit.

It must be a slash along the diagonal, however (Figure 4). The model will show four faces. There are two versions, the second allowing some presentational by-play. Use of paper money is recommended.

Using the Alternating Way Fold, flexing is not continuous, and two of the sides are just a little tricky to find. The designs are these: checkerboard; checkerboard; solid white; triangle in each corner, leaving a white square in the center.

Using the Same Way Fold, flexing is not possible without an additional move. The obvious way is to open up each side of the slit, and then move the two little flaps inside the slit to a position outside it. Then flexing is possible. Or, once you understand what is required, you can tuck the flaps properly as you make the valley folds. However, and best of all, the move can be done quite secretly in the course of doing the first flex. Being sure you have the right axis on which to fold the model in half; do the folding, but pull gently as you do it, and the little flaps will be adjusted automatically and properly, to be seen only at the very end of the flexing. Obviously, you can create some mischief by folding, flexing and unfolding, then challenging the knowledgeable paperfolder to duplicate the action. The flexing is continuous. The faces are these: solid white; two black (and six white) triangles opposite each other; two white triangles opposite each other, arranged differently from the previous two; a second solid white.

Square X-Slit

These models are made from squares with two intersecting slits making an X (Figure 5). Many versions of these are possible. Here are four of them. All have six faces.

Use of the Same Way Fold shows four faces continuously, and the other two easily, in the following manner: face 1, 2, 3, 4, 1, 5, 6, 4, 1, 2, etc. The faces are as follows: two black triangles opposite each other; two black triangles in a different arrangement; two white triangles arranged as in face 1; two white triangles, arranged as in face 2; solid white; solid black.

The Same Way Fold can also be used, with the inner flaps arranged differently: two opposite flaps creased one way, the other two creased the other way. The result flexes exactly the same as the one above, but with different designs: eight triangles of alternating color, circling about the center; two black triangles opposite each other; solid black; solid black; same as the second face; solid black. This does not appear to offer as much variation due to the repetition involved. However, the first face takes on a quite different appearance when flexed to the back of the model, as the eight triangles are rearranged in an interesting way — four triangles of alternating color circle about the center, with one triangle in each corner of the model.

Use of the Alternating Way Fold creates a puzzle version. Four faces are shown continually, and the other two are deceptively hard to find, although no movement to a ring is required. The designs are these: checkerboard square; triangles in each corner so the center forms a white square; four large triangles alternating in color; checkerboard square; solid black; solid white. (This is my favorite of these four because the designs are handsome and the two solid color faces are a little tricky to find.)

The Alternating Way Fold can be used with the inner flaps arranged differently, also. The designs on one pair of faces are different, two triangles on each face, but differing in both color and arrangement.

Possibilities

Other flexagons can be constructed from the X-slit approach. For example, I constructed one to reveal the maximum of variety in design. This uses the fact that all flexagons show some faces more than once. The second and third showings of the face reorient parts of it. So although my multiversion shows only six faces, these reveal seven different designs: solid color; checkerboard of four squares; one triangle (out of eight); four faces each with different arrangements of three triangles.

Additionally, the methods given above can be combined to create still different design possibilities. For example, when a half of a corner is removed, instead of all or none of it, the design changes. For another example, combine a half-cross base with a slash-slit, and use the Same Way Fold. And so on.

This manuscript was submitted for possible publication after the Gathering for Gardner in January of 1993, but written more than a year or two before that. One of the items was taught at the Gathering and appeared in the Martin Gardner Collection.

The Odyssey of the Figure Eight Puzzle

Stewart Coffin

I became established in the puzzle business in 1971, hand-crafting interlocking puzzles in fancy woods and selling them at craft shows as fast as I could turn them out. What I lacked from the start was a simple interesting puzzle that could be produced and sold at low cost to curious children (or their parents) who could not afford the fancy prices of those AP-ART sculptures. To fill this need, I turned my attention to topological puzzles, which almost by definiton do not require close tolerances and are just about the easiest to fabricate. This led to a couple of not very original or interesting string-and-bead puzzles (the Sleeper-Stoppers). I also tried to come up with something in string and wire that could be licensed for manufacture. I may have been inspired in this by a neat little puzzle known as Loony Loop that was enjoying commercial success at the time. The three rather unexciting puzzles that came out of this (Lamplighter, Liberty Bell, and Bottleneck) are described in the 1985 edition of *Puzzle Craft*, so I will not waste space on them here, except to describe their common design feature.

Take a foot or two of flexible wire and form a loop at both ends, thus:

Then bend the wire into some animated shape, with wire passing through the loops in various ways, such as in the very simple example below.

Finally, knot a sufficiently long loop of cord around some part of the wire and try to remove it (or try to put it back on after it has been removed). Unlike the simple example shown above, some of these puzzles can be quite baffling to solve.

Alas, none of these ideas enjoyed any success. (In a recent survey of my puzzle customers, they reported that topological puzzles in general were their least favorite of any puzzles I had produced.)

The story does not end there, however. For a slight variation of this, during an idle moment one day I formed some wire into the figure-eight shape shown below, knotted a cord around it, and tried to remove it.

I soon became convinced that this was impossible, but being a novice in the field of topology, I was at a loss for any sort of formal proof. I published this simply as a curiosity in a 1974 newsletter (later reprinted in *Puzzle Craft*). Some readers misread that purposely vague write-up and assumed that it must have a solution, which then left them utterly baffled as to finding it. Royce Lowe of Juneau, Alaska, decided to add the Figure Eight to the line of puzzles that he made and sold in his spare time. When some of his customers started asking for the solution, he begged me for help.

Next it appeared in a 1976 issue of a British magazine on puzzles and games. The puzzle editor made the surprising observation that it was topologically equivalent to the Double-Treble-Clef Puzzle made by Pentangle and therefore solvable, since the Pentangle puzzle was. But a careful check showed that the two were not quite equivalent.

To add to the confusion, somewhat to my surprise, the puzzle appeared in *Creative Puzzles of the World* by van Delft and Botermans (1978) with a farce of a convoluted "solution" thrown in for added amusement. More recently, I received a seven-page document from someone in Japan, full of diagrams and such, purporting to prove that the puzzle was unsolvable. The proof appeared to be a rather complicated, and I did not spend a lot of time trying to digest it. Over the intervening years, I had continued to received numerous requests for the solution, and by that time I was rather tired of the whole thing.

When it comes to puzzles, it is often the simplest things that prove to have the greatest appeal, probably not even realized at the start. Whoever would have guessed that this little bent piece of scrap wire and loop of string would launch itself on an odyssey that would carry it around the world? I wonder if this will be the final chapter in the life of the infamous Figure Eight Puzzle. Or will it mischievously rise again, perhaps disguised in another form, as topological puzzles so often do?

Metagrobolizers of Wire

Rick Irby

Many people love the challenge of solving a good puzzle. In fact, those who like puzzles generally like to solve just about any problem. Be it a paradox, a mathematical problem, magic, or a puzzle, the search for answers drives many of us on. Unlike magic or illusions with misdirection and hidden mechanisms, mechanical puzzles are an open book, with everything visible, all parts exposed ready for minute examination and scrutiny. In spite of this, the solutions can elude even the sharpest and quickest minds of every discipline.

Puzzles can go beyond an understanding of the problem and its solution, and here is where the separation between the common puzzler and (to borrow a phrase from Nob Yoshigahara) a "certifiable puzzle crazy" lies. The majority of mechanical puzzle solvers take the puzzle apart through a series of random moves with no thought given to the fact that this way they have only half-solved the problem. The random-move method will suffice for easy to medium puzzles but will do little or no good for solving the more difficult ones. A "puzzle crazy," on the other hand, will analyze the problem with logic and stratagem, then reason out the solution to include returning the parts to their original starting position. Regardless of one's ability to solve them, puzzles entertain, mystify and educate, and the search for puzzling challenges will undoubtedly continue.

My own interest in puzzles began in early childhood, with the small packaged and manufactured wire puzzles available at the local 5 & 10-cents store. Although entertaining, they were never quite enough of a challenge to satisfy my hunger. Somewhere in the back of my mind I knew I could come up with better puzzles than were currently and commercially available. About twenty years after my introduction to those first little wire teasers, a back injury from an auto accident and lots of encouragement from puzzle collectors brought the following and many other puzzle ideas to fruition.

Wire disentanglement puzzles are topological in nature and can vary widely in both difficulty and complexity of design. Wire lends itself very easily to topological problems because of its inherent nature to be readily and easily formed into whatever permanent shapes may be necessary to present a concept.

I am frequently asked to explain the thought process involved in coming up with a new puzzle. Unfortunately, I really can't answer. I neither know or understand the process of any creativity than to say that it just happens. It is my suspicion that the subconscious mind is constantly at work attempting to fit pieces of countless puzzles together; sometimes it succeeds! If you devote your mind to something, either you become good at it or you are devoting your mind to the wrong thing. A couple of examples of ideas that have "popped" out of my mind at various times are explained and illustrated below.

Many thanks must go to Martin Gardner as an inspiration to the millions enlightened by his myriad works. Thanks also to Tom Rodgers for his support of my work, for asking me to participate in "Puzzles: Beyond the Borders of the Mind," and for presenting me with the opportunity to meet Martin Gardner.

The Bermuda Triangle Puzzle

Knowing the fascination that many people have with the somewhat mysterious and as-yet-unexplained disappearances of various airplanes and ships in the area known as the Bermuda Triangle and the Devil's Triangle made naming this puzzle relatively easy. Often it is easier to come up with and develop a new puzzle idea than to give it a good and catchy name. This puzzle idea came to me as I was driving to San Francisco to sell my puzzles at Fisherman's Wharf, in 1971 or 1972.

The Bermuda Triangle Puzzle

The object of the Bermuda Triangle is to save the UFO that is trapped in the puzzle, the UFO being a ring with an abstract shape mated to it. The puzzle is generally set up with the ring around the Bermuda Triangle, which is a triangular piece. The triangle can be moved over the entirety of the larger configuration, taking the UFO with it as it moves. There are several places where the UFO may be separated from the triangle but only one place where the separation will allow the solution to be executed. Most of the large configuration to which the triangle is attached is there simply to bewilder the would-be solver. The solutions to many puzzles can be elusive until the puzzle has been manipulated many times; although moderately difficult for the average puzzler to solve initially, this one is relatively easy to remember once the solution has been seen. The Bermuda Triangle rates about a medium level of difficulty.

The Nightmare Puzzle

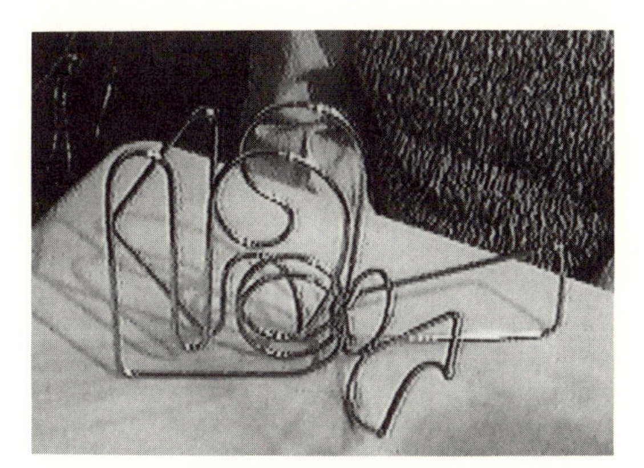

The Nightmare Puzzle

The Nightmare puzzle was conceived as Johnny Carson was delivering his monologue during the Tonight Show one night in 1984. As with most of my puzzle ideas, this one came to me fully formed and complete. I keep tools and wire handy for just such events having learned that three-dimensional ideas are difficult to decipher and duplicate from two-dimensional scribbling on a piece of paper. After making the prototype, as I sat playing with it, my wife joined me in critiquing my latest design. Sometime later one of us stated that we probably would have nightmares about it that night. We didn't have any nightmares about it, but the name stuck. The Nightmare has more than lived up to its name with a convoluted three-dimensional shape that exacts extreme effort and concentration from all who attempt it.

The Nightmare puzzle is made from one continuous strand of wire. There are two outer and two inner loops, with the wire ends making small rings that are wrapped around the wire in such a manner as to eliminate any usable ends. A cord encircles the two inner loops of the puzzle, and the object is to remove the cord completely from the puzzle. In addition to the difficulty in conceptualizing the convoluted shape of the puzzle, the flexibility of the cord allows one to make mistakes not possible with rigid pieces. Any wrong moves, not promptly corrected, quickly compound into a tangled mass of knots soon precluding any progression toward the solution. On a scale of 1 to 10, I rate the Nightmare an 8. The difficulty may be increased by, after the cord is removed, adding a ring to the cord that will not pass through either of the small end rings then attempting to replace the cord.

Beautiful But Wrong: The Floating Hourglass Puzzle

Scot Morris

The Beginning

One of the problems in Martin Gardner's August 1966 "Mathematical Games" column was The Floating Hourglass.

> An unusual toy is on sale at a Paris shop: a glass cylinder, filled with water, and at the top an hourglass floats. If the cylinder is inverted a curious thing happens; the hourglass remains at the bottom of the cylinder until a certain quantity of sand has flowed into its lower compartment. Then it rises slowly to the top. It seems impossible that a transfer of sand from top to bottom of the hourglass would have any effect on its overall buoyancy. Can you guess the simple *modus operandi*?

I gave the problem some thought but couldn't come up with any good theory. I assumed it had to do with some law that I had forgotten since high school. The next month, when the answer came, I was delighted. It was so simple, so absurdly obvious, that I not only *could* have thought of it myself, I *should* have. The effect was like seeing a good magic trick or hearing a good joke.

I read Martin Gardner's columns religiously, and corresponded with him occasionally from 1963 on, as a college student, as a graduate student, and then as an editor of *Psychology Today*. In 1978, in the months before a new science magazine, Omni, was to be launched, I had the pleasure of finally meeting the Master Explainer. I was going to write a column on "Games" so I made a pilgrimage to Hastings-on-Hudson to visit the Master. There on a shelf was the infamous Floating Hourglass itself. I could finally try out the curious toy I had read about so many years before. I turned it over and the hourglass stayed at the bottom, just like Martin said it would.

At the 1991 Puzzle Collectors Party in Los Angeles, I saw my second Floating Hourglass and knew I could finally write about it, since I only published puzzles in *Omni* that I knew my readers could find. Tim Rowett had brought one from England, made by Ray Bathke of London. I immediately ordered some. My September 1992 column introduced the Floating Hourglass 25 years after I first heard about it. I asked my clever readers to submit theories to explain it. The results appeared in the January 1993 issue.

When Martin allowed me to look through his files, I found a thick folder on the Hourglass, a treasure trove of letters and drawings. For years I have itched to tell this story, but no magazine article could possibly contain it. This book finally gives me the chance to tell the history of the Floating Hourglass Puzzle.

The First Theories

Piet Hein, the Danish sculptor/inventor (the Soma Cube, the Super Egg) and poet/artist (Grooks), had visited Martin in early 1966. Hein told him about a toy he had seen in the Paris airport. He didn't bring one back, but his description was clear enough; Gardner knew how it must work and wrote about it in his August and September columns without even having seen one. Martin's theory relied on friction between the glass and the cylinder, but Pien wasn't convinced. He thought the inside of the cylinder he had seen was too smooth to offer much resistance. He felt there must be something more, something to do with the falling sand.

Hein believed the impact of the sand grains hitting the bottom of the glass exerts a downward force, an 'effective change in mass': "The hourglass is heavier while the sand is falling," Hein wrote Gardner on September 16, 1966. "Imagine if the hourglass were opaque and you didn't know it were an hourglass at all. There it stands at the bottom and changes its weight!...What keeps the hourglass down is the falling of the sand, not the amount of sand that has fallen or is left. The hourglass rises not because there is little sand left in the top chamber, but because the rate of falling sand has decreased. This seems to solve the whole problem."

Note that he wrote all this after the September *Scientific American* was out. He knew of Gardner's "answer," but he also knew that Martin had never examined an actual hourglass sample. For him this meant that the final proof was not yet in. Until you could see and touch one, break it open or learn how it is made, all theories were valid contenders. In the absence of knowing the truth, the best criterion for a theory is its beauty. And Martin had already admitted that Piet's impact theory was beautiful.

Hein was obsessed with hourglasses. He drew a cartoon of himself on an elevator with Einstein, pondering an oversized hourglass. He designed an hourglass-powered perpetual motion machine and created a fantasy ocean full of bobbing hourglasses. Since the glasses change their weight whenever the sand falls inside, Hein issued a mock warning: When mailing hourglasses, don't weigh the package while they're running, or you'll have to pay a higher postage.

A Painful Paradigm Shift

Just a couple of days after writing the letter and drawing the cartoon, Hein had an agonizing experience in Milan, Italy. In a shop there he saw a double-glass: two cylinders side by side, a floating hourglass in one and a sunken hourglass in the other. When turned over, the glasses stay in place at first and then one rises while the other sinks. He knew immediately that his impact theory was doomed. A "sunken" hourglass that stays in place at the top of a tube can't be explained by sand grains falling in the opposite direction.

Hein tried the double-glass in the shop a few times, just enough to see that it worked. "That was all I wanted to know," he wrote "and all that was needed to make my intellectual headache come back much worse than the first time." Hein was forced to make a paradigm shift, and he found it painful. He knew that the sinking glass directly refuted his theory, but he didn't buy one to take home. "This is not a question of fumbling one's way to a solution, but of thinking," he explained. "I couldn't think of anything else but the principle. It really hurt."

He couldn't bear to abandon his pet idea *completely*. He rationalized in jest: "My theory works, with the exception that the sand was running downward in both hourglasses, I must admit, but it would be easier to explain the symmetry of the phenomenon if it were falling upward in one of them!"

Hein acknowledged Gardner's appreciation of beauty in a theory. "I am glad you think my explanation is beautiful," he wrote. "So do I, but let us be honest and not rate it any lower just because it is false. I admit, being false makes it less right. But it does not make it any less beautiful."

When Piet Hein left Milan on the afternoon of September 21, he was somber, "somewhat worried and depressed on behalf of my beautiful theory." Then, as the plane soared over Mont Blanc, Hein had a sudden crystallizing insight - "At an altitude of 8500 meters, I found myself in possession of the solution." He later wrote about the "Aha" experience. "When the Alps in the lite of the (superelliptic) setting sun and the growing, exactly half moon (D for dynamics) were left behind us and all was dark, there was a lull in my mind, a *tabula rasa*. And on that tablet lay, like a small fish on the center of a huge platter, the solution of the hourglass mystery. My dear Watson, it's very simple."

The Solution Is in the Solution

The key to the puzzle isn't *in* the glass but *around* it, Hein realized, not in the falling of the sand but in the flowing of the liquid. There must be two liquids of different specific weights that don't mix completely, and are indistinguishable in color and transparency.

When the cylinder is inverted, the heavier liquid is on top pushing down. Only when enough of it has seeped down does the glass begin to rise. The falling sand is just for misdirection, but what a clever ruse it is. The hourglass puzzle is caused by an hourglass effect, but the relevant displacement is in the liquid, not the sand. Hein was awed by the brilliance of it all, "I should like to meet the person who invented this effect and designed it so as to hide the solution so elegantly for us," he wrote on September 22.

Gardner's reply on October 5th was a gentle letdown: "You said you would like to meet the clever fellow who thought of the 'two-liquid' principle.

Well, all you have to do is shake your own hand. You are the inventor. The principle is simply too clever to be true."

The Hourglass Letters

While Hein was having his epiphany in Europe, Gardner was getting letters in response to the September issue. One man wrote that he owned one of the Paris cylinders, but reported that his glass sometimes floated and sometimes sank, perhaps depending on the temperature.

On September 6th, Albert Altman wrote from the U.S. Naval Ordnance Laboratory at Silver Spring, Maryland:

> Another solution to the hourglass science teaser is that the momentum carried by the falling sand causes the hourglass plus sand to weigh more than its static weight by an amount $\mu\sqrt{2gh}$, where μ is the rate of the flow of the sand, g the acceleration of gravity, and h the height through which the sand has fallen. The height decreases due to the buildup of sand on the bottom of the hourglass and at a critical value the net force on the hourglass acts upward and rises.

Gardner was beginning to wonder if his friction explanation told the whole story. Could temperature and sand impact also be factors? Would he have to print a correction? He replied to Altman on September 12th:

> I am embarrassed to admit that your explanation may be right. I have not yet seen the toy; having relied (unfortunately) on an account given to me by a friend who examined the toy in Paris, but did not bring one back with him. It is possible that one version of the toy works on the principle you mention, and the other on the principle I suggested, or perhaps still another one. In short, at this point I am hopelessly confused.

Hopelessly confused? Martin Gardner?! That is something that surely doesn't happen very often, and it didn't last long. A letter dated September 29 came from Walter P. Reid, also from the U.S. Naval Ordnance Lab. "I am writing to put your mind at ease (on the impact theory posed by Altman), and to suggest that you not publish a correction. I am sure that your explanation was correct." Reid went on to show mathematically how the impact of sand hitting the glass' bottom is exactly balanced by the loss of the sand's mass while it is in free fall. Reid later adapted his letter for publication and his short article "Weight of an Hourglass" appeared in *American Journal of Physics,* **35**(4), April, 1967. This remains, to my knowledge, the only scientific writing on the subject. Gardner cites it in the hourglass puzzle reprinted in *Mathematical Circus.*

Hein and the Horse's Mouth

All was settled for a few months, but then in the spring of 1967 the glass rose again. Hein wrote that he went back to the shop in Paris where he first saw the glass, and tracked down the maker. He turned out to be a Czechoslovakian glassblower named Willy Dietermann, and he confirmed Hein's two-liquid theory. When Hein asked about the liquid, Mr. Dietermann explained it was 90 to 95% water and 5 to 10% a combination of alcohol and formic acid.

Gardner wrote to Dietermann but received no reply. He decided he had to write a correction after all. Set into type and slated to appear in a summer 1967 column was this recreation of the moment of truth: "When Hein stated his two-liquid theory, the inventor staggered as if struck. It was the first time his secret had been guessed: two liquids of different density, but which do not separate completely, so there is no visible division between the two. A slight difference in refractive power is concealed by the curvature of the glass cylinder. The glasses do tip to the side when inverted, but that is just a holding operation until the liquids have had time to change places. The main principle is the liquid hourglass effect, completely invisible, which eventually sends one hourglass up, the other down."

Proof Positive

By this time, Gardner had his own hourglass tube. He loaned it to C.L. "Red" Stong, the original "Amateur Scientist" columnist of *Scientific American.* In his lab, Stong shone polarized light through the cylinder and found that the refractive index of the liquid did not change from top to bottom. He concluded the liquid inside was homogenous. He also reported it had a freezing point of eight degrees below zero.

Meanwhile, Gardner was playing amateur scientist himself. He bought a cheap egg timer, freed the glass from the wooden frame, and found a transparent cylinder into which it would fit. "I filled the cylinder with water," he wrote to Hein, "then I wrapped copper wire around the middle of the hourglass. By snipping off the ends of the wire I was able to make its weight such that it slowly rises in the cylinder. It worked just like the big version."

Sanity was restored and the correction was never printed. We may never know what Hein got from the "horse's mouth." Perhaps Dieterman was leading him along, or perhaps there was an honest misunderstanding about the use of different liquids. There certainly was plenty of room for misinterpretations. The Dane and the Czech had to speak German because Dieterman knew no Danish or English, and his German was better than his French, and Hein's French was worse than his German.

Final Thoughts

What fascinated me about the hourglass puzzle was how it led a mind like Piet Hein's to come up with such brilliantly incorrect theories. They may have been wrong, but they were creative products of human thought, and deserved to be prized for that alone. Let others measure a refractive index or a freezing point, Hein wanted to think the problem through. He wanted to search for alternate, beautiful explanations. He wanted to expand his "perpendicular thinking."

I received over 1200 letters about the hourglass after publishing the puzzle in *Omni*. As I read them, sorting them into different piles, I found the largest single category was always the "correct" theory. This proportion stays at about 40% with each new batch of mail. The other 60% broke down into about 15 different theories.

About 40% of those readers with incorrect answers cited heat as a factor in the hourglass's behavior. Falling sand generates heat, they said. Some argued that this warms the surrounding liquid so the hourglass stays down until the liquid cools again; others, that the hourglass floates up with the pocket of warm liquid surrounding the glass's neck. But most in this category thought the heat warms the *air* in the glass, making it expand slightly and then rise.

More than 50 readers thought that the hourglass was flexible. Some reasoned that when the sand presses down from the top, the hourglass widens and wedges itself into the cylinder. Others decided that the hourglass is flexible only at the ends. "The top and bottom of the hourglass are so thin as to sag under the weight of the sand," wrote B. G. of Los Altos Hills, California. "When enough sand falls into the bottom chamber, it 'bubbles' the bottom end out, increasing the hourglass's volume," reasoned D. Q. of Richmond Hill, Ontario, Canada.

Many correspondents blamed the "impact" of falling sand for keeping the glass down. Some even used mathematical formulas to show how much force a sand grain exerted, first on the bottom of the glass and later on a mound of other sand grains. The theory may be correct, but the calculations have to consider the amount of time each grain of sand is falling and weightless; the two tiny opposing forces exactly cancel each other out. Movements within the system don't alter the weight of the system.

Another line of reasoning put the emphasis on the liquid: "The solution is in the solution!" wrote H. W. of Coweta, Oklahoma. If the liquid is naturally cooler and denser at the bottom, then the denser liquid is at the top when the tube is turned over. It eventually seeps down below the hourglass and buoys it up.

A surprising number thought it was all an illusion. "It just takes a long time for the hourglass to get started," perhaps because the liquid is very viscous, wrote one reader. "The hourglass, *as a system,* is rising from the moment the column is inverted," argued P. T. of Glendale, California. They concentrated on the air bubble that constantly rises, first in the hourglass and then in the tube.

Many believed the air at the *top* of the hourglass lifts it to the top of the tube. "When enough air reaches the top chamber and exerts its pressure there, the hourglass begins to rise," wrote T. H. of Chapel Hill, North Carolina. About 4% of those who wrote in thought that the shape of the hourglass affected its buoyancy. When the air is in the bottom half, the water below the glass can push up only on the circular end of the glass. When the air rises to the top half, water can push up all around the inverted cone, a greater surface area. "It's the same principle that causes a snow cone to pop out of its cup when you squeeze the bottom," explained D. A. N. of Tillamook, Oregon.

I promised copies of the book *Omni Games* to the five "most interesting" entries. Correct answers to this puzzle aren't very interesting because they're all virtually alike. Therefore, I awarded books for *incorrect* answers only:

1. Pete Roche of Chicago foresaw this and sent both a correct theory and this incorrect alternate: "The hourglass has a small clasping mechanism at each end. The momentum of rising provides enough energy to engage the mechanism as it reaches the top of the cylinder, while the weight of the falling sand is required to release the mechanism."
2. John J. Gagne of Eglin Air Force Base, Florida, supposed that sand blocked the hole, and air in the bottom of the hourglass became compressed until "a jet of air shoots into the top half of the hourglass, imparting just enough lift to overcome inertia and start the glass moving up."
3. "The hourglass is composed of a flexible material such as Nalgene," wrote Timothy T. Dinger, Ph.D., and Daniel E. Edelstein, Ph.D.

The two share a prize for having the confidence to submit their incorrect theory on company stationery: IBM's Thomas J. Watson Research Center in Yorktown Heights, New York.

4. Cliff Oberg of Clarkdale, Arizona, theorized that the tube's caps are hollow and that the fluid must flow from the tube through a hole into the cap below before the glass can rise.
5. Finally, a theory about the density gradient of the liquid was signed "Bob Saville, Physics Teacher, Shoreham-Wading River High School, Shoreham, New York." He added, "P.S. If this is wrong, then my name is John Holzapfel and I teach chemistry." The fifth book, therefore, goes to Holzapfel.

> The "correct" answer, as Martin Gardner wrote it in 1966: When the sand is at the top of the hourglass, a high center of gravity tips the hourglass to one side. The resulting friction against the side of the cylinder is sufficient to keep it at the bottom of the cylinder. After enough sand has flowed down to make the hourglass float upright, the loss of friction enables it to rise.

Cube Puzzles

Jeremiah Farrell

Binomial Puzzle

A new, rather amusing, combinatorial puzzle can be constructed by acquiring twenty-seven uniform cubes of size $a \times a \times a$ and gluing them together to form the following eight pieces:
One "Smally" of size $a \times a \times a$ (i.e., a single cube).
One "Biggie" of size $b \times b \times b$ (where $b = 2a$).
Three "flats" of size $a \times b \times b$.
Three "longs" of size $a \times a \times b$.
Now color each piece with six colors according to the scheme in Figure 1.

The problem then is to arrange the eight pieces into a cube so that opposite faces have the same color.

This puzzle, if colored as above, has exactly two solutions – each with a different set of three colors. With proper insight it can be solved in a few minutes. Without this insight it typically takes several hours to arrive at a solution if one can be found at all. Before proceeding with this discussion the reader is urged to build a puzzle and attempt its solution.

Most Martin Gardner aficionados will recognize that the eight pieces model the situation represented by the binomial expansion

$$(a + b)^3 = a^3 + 3a^2b + 3ab^2 + b^3 \tag{1}$$

Gardner calls such models of mathematical theorems "look-see" proofs. In fact he has recommended that every teacher of algebra construct a set of eight pieces for classroom use. Usually there is an "aha" reaction when students see that a cube can actually be constructed from the pieces. For more advanced students it would be reasonable to ask in how many essentially different ways can the cube be constructed from the eight (uncolored) pieces. This will depend on just what is meant by the words "essentially different" but one interpretation could be to orient the cube to sit in the positive octant

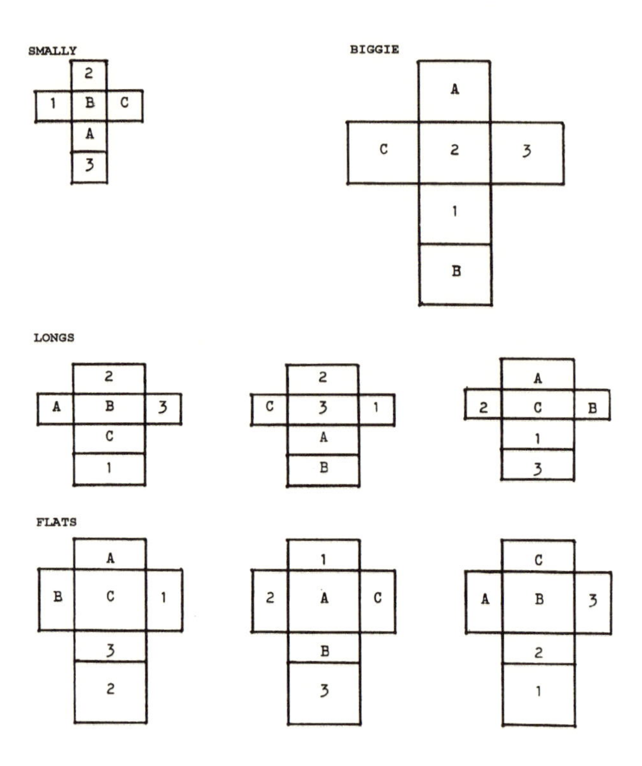

Figure 1. A, B, C, 1, 2, and 3 are any six distinctive colors.

in space with Biggie always occupying the corner (0, 0, 0). There are 93 solutions in this case. If, additionally, each of the 48 faces of the pieces is colored with a unique color, then there are $(93)(8^8)(3!)^2 = 56{,}170{,}119{,}168$ distinct ways of constructing the cube.

The combinatorial puzzle uses only six colors in its construction, and the solver has the additional clue that a face is all of one color; but there still remain a great many cases that are nearly right. Trial and error is not a very fruitful way of trying to solve this puzzle.

Solution Hints. These hints are to be regarded as progressive. That is, after reading a hint, try again to solve the puzzle. If you cannot, proceed to the next hint.

1. Call the six colors A, B, C, 1, 2, and 3. A, B, and C will turn out to be a solution set. Notice that the eight pieces must form the eight corners of the completed cube – one piece for each corner. Therefore, each piece must have on it a corner with the three colors

A, B, and C in some order. (It is a fact, but not necessary for the solution, that four pieces will have the counterclockwise order A-B-C and the other four the order A-C-B.)

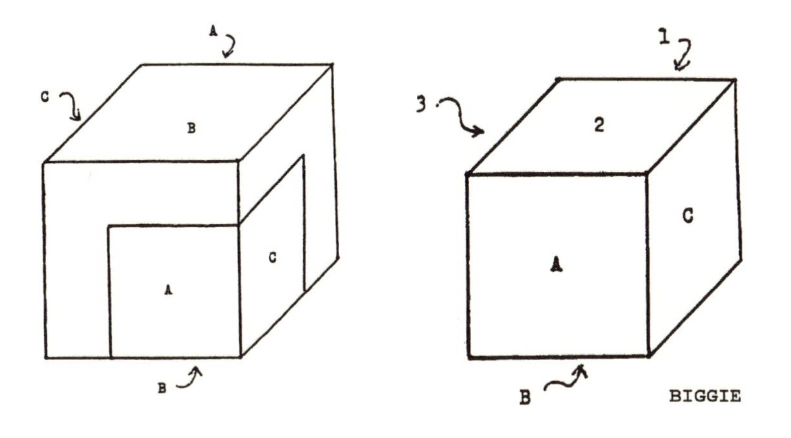

Figure 2.

2. To determine the colors A, B, and C, take any piece (Biggie is a good choice), and notice that the three pairs of colors A-1, B-2, and C-3 oppose each other. This means that A and 1 cannot appear together in a solution. Likewise B and 2 cannot, nor can C and 3. Of the 20 possible sets of three colors (out of six) we are left with only eight possibilities: A-B-C, A-B-3, A-2-C, A-2-3, 1-B-C, 1-B-3, 1-2-C, and 1-2-3.
3. Since it is also true that, on any other piece, colors that oppose each other cannot appear in the same solution, we may choose another piece, say a flat, and use it to reduce further the possible solution colors. For instance, it may happen that on that flat, 2 opposes A. We would then know to eliminate anything with A and 2. This would force A and B to be together. One or two more tests with other pieces lead to A-B-C (or 1-2-3) as a candidate for a solution. When eliminating possibilities, it is convenient to turn the three flats to allowed colors, using Biggie as a guide.
4. Place Biggie as in the diagram so that A-B-C is a corner. Place all three flats so that A, B, and C are showing. These three flats must cover all or part of the colors 1, 2, and 3 of Biggie. Of course, keep A opposite A, etc. It is easy to visualize exactly where a particular

flat must go by looking at its A–B–C corner. Then place the longs and, finally, Smally.

The solver will notice that the completed cube has a "fault-free" property. That is, no seam runs completely through the cube in any direction and therefore no rotation of any part of the cube is possible. This will assure that only one solution is attainable with the colors A–B–C (the proof is left to the reader). There is another solution to this puzzle using the colors 1–2–3. If only one solution is desired, take any piece and interchange two of the colors A, B, and C (or two of 1, 2, and 3). This will change the parity on one corner of the cube so that a solution is impossible using that set of three colors.

We like to have three Smallys prepared: one that yields two solutions, one that gives only one solution, and one to slip in when our enemies try the puzzle that is colored so that *no* solution is possible!

Magic Die

Figure 3 shows the schematic for a Magic Die. The Magic Die has the amazing property that the sum of any row, column, main diagonal (upper left to lower right), or off-diagonal around all four lateral faces is always 42.

Figure 3.

You can construct a Magic Die puzzle by taking twenty-seven dice and gluing them together into the eight pieces of our combinatorial cube puzzle—being sure the dice conform to the layout shown in Figure 3. There will be only one solution to this puzzle (with magic constant 42), and, even with the schematic as a guide, it will be extremely difficult to find. (Alternatively, you can copy Figure 3 and paste it onto heavy paper to make a permanent Magic Die.)

The Nine Color Puzzle

Sivy Farhi

The nine color puzzle consists of a tricube, with each cube a different color, and twelve different dicubes with each cube of a dicube a different color. Altogether there are three cubes each of nine different colors. The object of the puzzle is to assemble the pieces into a cube with all nine colors displayed on the six faces. A typical set is shown below in Figure 1, where each number represents a color. Typical solutions are shown at the end of this paper.

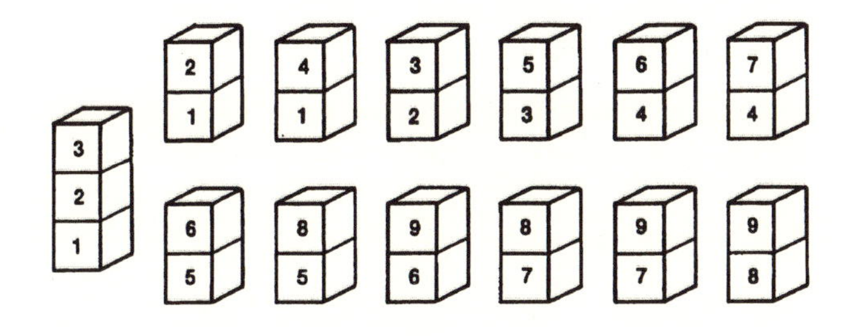

Figure 1.

According to Reference 1, the nine color puzzle was first introduced during 1973, in Canada, as "Kolor Kraze." I learned of the puzzle in 1977 from Reference 2, and, using the name "Nonahuebes," included it among the puzzles I produced in New Zealand under the aegis of Pentacube Puzzles, Ltd. I was intrigued by the puzzle's characteristics but did not, until recently, analyze it thoroughly.

This puzzle is of particular interest because it is of simple construction, has an easily understood objective, has numerous variations and solutions, is a manipulative puzzle, and requires logical decision-making. Included in its analysis are the concepts of transformations, parity, backtracking, combinatorics, ordered pairs, sets, and isomorphisms. A complete analysis requires the use of a computer, but after reading this article it should not be a difficult exercise for a moderately proficient programmer to verify the results obtained.

The first observation noted is that the cube can be assembled with the tricube along the edge or through the center of the cube, but not in the center of a face. A proof is given in Section 2.

The second observation is that for a cube to be assembled with the nine colors on each face, none of the nine planes may contain two cubes of the same color. A proof is given in Section 3. Thus, when two cubes of the same color are in position, the location of the third cube is determined. Since during the course of trying to solve this puzzle a conflict often occurs, this rule then tells the experimenter to backtrack.

The third observation, obtained after some experimentation, is that there are numerous solutions, some with the tricube on the edge and some with the tricube through the center of the cube. The question then arises: How many solutions are there?

Most intriguing about this puzzle, and the most difficult aspect to investigate was the fourth observation: the color combinations need not be as shown in Figure 1. The previous question now becomes more interesting. How many solutions are there for each color combination? And how many color combinations are there?

This paper addresses these last two questions. The individual dicubes can be colored thirty-six different ways. Twelve of these can be selected in 1,251,677,700 (36!/12!24!) different ways of which, as determined in Section 4, only 133,105 of these combinations meet the three-cubes-of-each-color requirement. Since one person can assign the fourth and fifth colors to orange and red and another person to red and orange, it becomes apparent that most of the 133,105 combinations are isomorphisms. It becomes necessary, then, to separate these cases into disjoint sets of isomorphisms. Fortunately, as is shown in Section 5, only thirty dicubes are required, and only 10,691 combinations need to be sorted into disjoint sets.

The method of sorting the 10,691 combinations into 148 disjoint sets is described in Section 6, in which the number of isomorphisms using the set of thirty-six dicubes is also determined.

The final part of this analysis was to determine the number of puzzle solutions for each disjoint set of dicube combinations. This was achieved through the use of a computer program; the results are shown in Table 1.

Table 1.

Isomorphisms				Number of Solutions				
		dicubes		straight		L shape		
No.	Name	36	30	Edge	Center	Corner	Edge	Center
1	.:Z[yz	60	60	0	0	0	0	0
2	.:Zppp	10	10	0	0	0	0	0
3	.COeyz	360	180	8	0	16	16	0
4	.DOeyz	720	312	20	0	12	0	4
5	.DZ[yz	1439	624	96	4	80	6	15
6	.DZpez	720	312	198	12	83	10	16
7	.DZppp	360	156	428	24	208	8	36
8	/90eyz	180	60	16	0	38	32	0
9	/:0eyz	360	120	80	4	24	4	4
10	/:Z[yz	720	240	182	8	50	16	24
11	/:Zpez	360	120	124	4	38	44	16
12	/:Zppp	180	60	408	0	186	32	64
13	/COeyz	360	72	0	0	6	4	0
14	/DXeyz	720	132	18	0	4	2	0
15	/DZ[yz	720	132	72	0	32	12	0
16	/DZepz	1439	264	41	0	40	9	0
17	/DZppp	360	66	80	0	117	28	0
18	/MNeyz	60	8	0	0	0	0	0
19	/MOeyz	720	96	8	0	34	8	16
20	/MXeyz	720	84	32	0	30	12	12
21	/MYeyz	720	84	26	0	18	6	4
22	/MZ[yz	1439	168	62	2	70	21	23
23	/MZepz	1439	168	41	3	35	15	16
24	/MZpez	1439	168	51	4	57	17	20
25	/MZppp	720	84	68	4	73	26	14
26	/NXeyz	720	96	8	0	36	12	12
27	/NYeyz	1439	192	40	2	64	17	24
28	/NZmyz	720	96	8	0	10	2	0
29	/NZnpz	1439	192	53	2	53	11	17
30	/NZpez	720	96	36	4	48	24	32
31	/NZpfp	1439	192	10	0	49	34	23
32	/Ob[yz	1439	168	25	1	81	24	16
33	/Obmyz	1439	168	98	5	65	18	19
34	/Obofz	720	84	6	0	18	6	8
35	/Obpez	1439	168	124	3	53	8	17

Isomorphisms		Number of Solutions						
		dicubes		straight		L shape		
No.	Name	36	30	Edge	Center	Corner	Edge	Center
36	/Obppp	720	84	38	2	104	46	20
37	/Oceyz	720	84	18	0	46	6	8
38	/Ocmyz	1439	168	34	3	27	12	2
39	/Ocnpz	1439	168	126	4	117	16	19
40	/Ocnyz	1439	168	40	1	36	11	11
41	/Ocnzp	1439	168	49	3	45	24	9
42	/Ocoxz	1439	168	112	5	14	3	2
43	/Ocozp	1439	168	18	1	35	28	9
44	/Oewez	720	84	32	2	38	48	4
45	/Oewnz	1439	168	52	3	20	15	1
46	/Oewpp	1439	168	17	6	38	12	4
47	/Oeyxp	720	84	36	0	12	6	4
48	9DEeyz	180	36	0	0	16	0	0
49	9DNeyz	30	6	0	0	0	0	0
50	9DOeyz	360	72	0	0	32	12	4
51	9DXeyz	720	108	10	0	0	0	0
52	9DYeyz	720	108	22	2	38	18	6
53	9DZ[yz	1439	216	69	3	32	17	2
54	9DZepz	1439	216	71	7	32	16	7
55	9DZpez	1439	216	55	6	52	13	5
56	9DZppp	720	108	54	12	72	18	20
57	9ODeyz	360	36	4	0	4	0	0
58	9OF[yz	360	36	76	0	72	8	4
59	9OFepz	720	72	82	0	45	0	6
60	9OFppp	180	18	208	0	134	8	0
61	9Ob[yz	720	72	54	4	41	12	2
62	9Obepz	1439	144	145	7	48	13	5
63	9Obmyz	1439	144	87	2	31	8	0
64	9Obnpz	720	72	2	0	18	4	0
65	9Obofz	720	72	68	4	28	2	0
66	9Obpez	720	72	22	2	25	4	0
67	9Obpfp	720	72	174	6	71	10	0
68	9Obppp	1439	144	30	3	26	5	4
69	9Odmpz	720	72	186	6	54	10	2
70	9Odmzp	360	36	16	0	6	12	0
71	9Odoyz	360	36	76	8	66	16	4
72	9Oewpp	360	36	40	0	16	4	0
73	DDDeyz	15	1	0	0	0	0	0

Isomorphisms				Number of Solutions				
		dicubes		straight		L shape		
No.	Name	36	30	Edge	Center	Corner	Edge	Center
74	DDEeyz	720	36	2	2	2	2	0
75	DDZQyz	180	7	0	0	0	0	0
76	DDZ[yz	1439	56	1	3	5	0	1
77	DDZofz	360	14	4	0	2	0	2
78	DDZpez	720	28	0	4	4	0	0
79	DDZppp	360	14	0	12	6	0	4
80	DEDeyz	360	12	4	4	4	2	0
81	DEF[yz	1439	48	10	3	15	3	0
82	DEFepz	720	24	18	6	16	0	0
83	DEFppp	360	12	0	20	7	6	0
84	DEOeyz	720	24	0	1	2	0	2
85	DEPQyz	720	16	6	0	0	0	0
86	DEP[yz	1439	32	1	0	0	0	0
87	DEPepz	1439	32	8	3	3	1	2
88	DEPmyz	1439	32	11	2	1	0	1
89	DEPnpz	1439	32	23	5	3	0	1
90	DEPofz	1439	32	13	3	10	1	1
91	DEPpez	1439	32	17	2	17	1	3
92	DEPpfp	1439	32	2	1	0	0	0
93	DEPppp	1439	32	12	8	2	1	0
94	DEZQyz	1439	48	23	5	6	0	1
95	DEZ[yz	360	12	0	0	0	0	0
96	DEZeyz	1439	48	4	6	4	0	1
97	DEZnpz	720	24	26	2	4	0	0
98	DEZnyz	720	24	14	0	14	0	1
99	DEZnzp	720	24	4	4	4	0	2
100	DEZozp	720	24	10	14	6	0	3
101	DEeOyz	720	16	6	2	0	0	0
102	DEePpz	1439	32	17	0	0	0	0
103	DEeRez	720	16	44	0	0	0	0
104	DEeRfp	1439	32	16	0	0	0	0
105	DEeYyz	720	16	0	0	0	0	0
106	DEeZpz	720	16	22	0	2	0	0
107	DEeZyz	1439	32	8	3	6	0	0
108	DEeZzp	1439	32	4	0	0	0	0
109	DEecpz	1439	32	54	5	9	0	0
110	DEecyz	1439	32	39	3	48	3	0
111	DEeczp	1439	32	18	1	10	2	0

Isomorphisms				Number of Solutions				
		dicubes		straight		L shape		
No.	Name	36	30	Edge	Center	Corner	Edge	Center
112	DEedyz	1439	32	13	1	11	2	0
113	DEeexz	1439	32	28	0	23	2	0
114	DEeeyz	1439	32	15	1	2	0	0
115	DEewez	1439	32	27	2	8	0	3
116	DEewfp	1439	32	36	0	1	0	0
117	DEewxp	1439	32	34	1	2	0	1
118	DEexnz	1439	32	33	1	10	0	1
119	DEexxp	1439	32	31	1	47	1	1
120	DEexyy	1439	32	47	0	3	1	1
121	DZDQyz	90	1	0	0	0	0	0
122	DZD[yz	720	8	2	2	6	0	0
123	DZDofz	180	2	8	0	0	0	0
124	DZDpez	360	4	0	0	12	1	0
125	DZDppp	180	2	0	8	6	0	2
126	DZGOyz	1439	16	28	0	17	1	1
127	DZDQfz	720	8	4	0	2	0	0
128	DZGRez	1439	16	20	0	11	0	0
129	DZGRpp	720	8	62	0	74	4	2
130	DZGYyz	1439	16	15	3	9	2	0
131	DZGZpz	1439	16	39	3	9	0	0
132	DZGZzp	720	8	4	2	6	0	0
133	DZG[xz	720	8	26	0	15	0	0
134	DZGwez	1439	16	29	2	2	0	0
135	DZGwnz	1439	16	36	0	8	0	0
136	DZGwpp	720	8	4	1	3	0	0
137	DZGwxp	720	8	45	0	15	0	0
138	DZpO[z	180	2	0	0	2	0	0
139	DZpOez	360	4	25	0	3	0	0
140	DZpOnz	720	8	200	6	22	0	1
141	DZzOpp	1439	16	183	4	51	1	1
142	DZpQnz	360	4	8	0	0	0	0
143	DZpQxp	720	8	129	2	10	1	0
144	DZpZpp	720	8	172	6	61	2	0
145	DZp[\p	360	4	65	14	16	1	4
146	DZp[fp	360	4	24	6	12	4	1
147	DZp[fy	360	4	21	0	4	0	0
148	Dzp[pf	720	8	137	0	69	1	1

1. Additional Comments

The Kolor Kraze puzzle shown in Reference 1 and the nine color puzzle shown in Reference 2 are isomorphic, thus leading me to believe that they originated from the same source.

It is not necessary that the tricube be straight, and, as mentioned in Section 2, an irregular tricube may have three locations. The number of solutions for these cases are included in Table 1.

There are two rather startling phenomena shown in Table 1. While some color combinations within the set of thirty dicubes have up to 720 isomorphisms, there are two color combinations (Numbers 73 and 121) which, no matter how the colors are swapped, map onto themselves. Verifying this would be an interesting exercise for the reader. Stranger yet, these two cases have no puzzle solutions. Is this a coincidence?

2. Permissible Locations of the Tricube as Determined by Parity Restrictions

1	2	3
4	5	6
7	8	9

10	11	12
13	14	15
16	17	18

19	20	21
22	23	24
25	26	27

Figure 2.

The three layers of the cube are checkered as shown in Figure 2, where odd numbers represent one state and even numbers the other. Each dicube when similarly checkered has one even and one odd number. Since the cube has fourteen odd numbers and thirteen even numbers, the tricube must fill two odd numbered spaces.

There are only two possibilities. The tricube location filling spaces 1, 10, and 19 is referred to as an "edge case" while the tricube location in spaces 5, 14, and 23 is referred to as a "center case." Other possible locations are reflections or rotations of these and are not considered as separate puzzles solutions.

If an irregular tricube is used, there are three possible locations. The corner case fills spaces 1, 2, and 11; the edge case fills spaces 2, 5, and 11; the center case fills spaces 5, 11, and 14.

3. The "No Two in the Same Plane" Rule

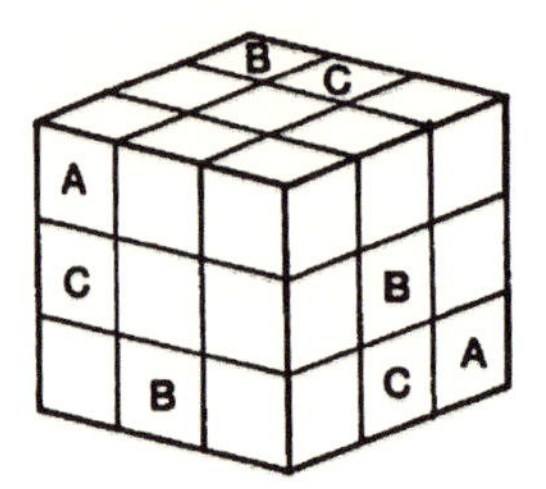

Figure 3.

One cube is centrally located and thus not visible. The only way the remaining two cubes of that color can be displayed on all six faces is for them to be located on opposite corners as shown in Figure 3 by the letter "A."

The remaining six corners must have cubes of different colors. Since the corner colors are displayed on three faces, the remaining two cubes must be located on an edge, accounting for two faces, and on a centerface such as shown by the letter "B." The remaining two colors must all be on edges, such as shown by the letter "C."

In each case, cubes of the same color are never in the same plane.

4. Determination of the 133,105 Possible Dicube Color Combinations

The tricube colors are defined as colors 1, 2, and 3. The two dicubes with color 1 may use any two colors in the first row of the following table. If color 2 is selected from the first row, then a third dicube with color 2 and any color from the second row is selected. If color 2 is not selected from the first row then color 2 is combined with any two colors from the second row for the third and fourth dicubes.

If none of the previous selections include color 3, then color 3 is combined with two colors from the third row. If color 3 has been selected once, then one color is selected from the third row and if color 3 has been selected twice, then none are selected from the third row.

First Color	Second Selection							
1	2	3	4	5	6	7	8	9
2		3	4	5	6	7	8	9
3			4	5	6	7	8	9
4				5	6	7	8	9
5					6	7	8	9
6						7	8	9
7							8	9
8								9

The process continues: color 4 is combined with colors in the fourth row such that there are three cubes with color 4, color 5 is combined with colors in the fifth row such that there are three cubes with color 5, etc.

Using this algorithm, a computer program established 133,105 possible color combinations.

5. Reduction of the Number of Isomorphisms

The table in Section 4 is simplified by restricting the selection in the first two rows as shown below.

Color	Selection					
1	2	3	4	5		
2		3	4	5	6	7

This can be justified by defining the colors associated with color 1 that are not 2 or 3 as colors 4 and 5. Similarly if color 2 is not associated with colors 3, 4, or 5, the two new colors are identified as colors 6 and 7. This results in requiring only thirty cubes and, using the same algorithm as before, only 10,691 color combinations.

6. Identification and Separation into Disjoint Sets

The color combinations in Figure 1 use dicubes with color (1, 2), (1, 4), (2, 3), (3, 5), (4, 6), (4, 7), (5, 6), (5, 8), (6, 9), (7, 8), (7, 9) and (8, 9).

Identification of the second color of these ordered pairs is sufficient to define all the colors. For example, given the second numbers 2, 4, 3, 5, 6, 7, 6, 8, 9, 8, 9, and 9 one can logically deduce the first colors by the three cubes of each color requirement and the restriction that the second number be greater than the first.

Identification was further simplified by placing these numbers in pairs (see Note below), i.e., (24), (35), (67), (68), (98), and (99), and then assigning a printable character to each, by adding twenty-three to each and using the ASCII character set codes (ASCII stands for American Standard Code for Information Interchange). Thus this color combination is identified as /:Z[yz.

The use of this six-character identification greatly simplified the process of sorting the 10,691 cases into 148 disjoint sets of isomorphisms.

Decoding a name is quite simple. For example, the first entry in Table 1 is .:Z[yz. The ASCII codes for these characters are: 46, 58, 90, 91, 121, and 122. Subtracting twenty-three results in: 23, 35, 67, 68, 98, and 99. It is then a simple process to recognize that this represents the dicubes colored (1, 2), (1, 3), (2, 3), (4, 5), (4, 6), (4, 7), (5, 6), (5, 8), (6, 9), (7, 8), (7, 9), and (8, 9).

Any permissible interchange of colors is an isomorphism. Color 2, which is in the center of the tricube, cannot be interchanged with other colors. Colors 1 and 3, on the ends of the tricube, can be swapped with each other, but not with any of the other colors. The remaining six colors may be interchanged in numerous ways: two at a time, three at a time, four at a time, including pairs of two at a time, five at a time including three at a time with two at a time, and six at a time including a triplet of two at a time, pairs of three at a time and four at a time with two at a time. Not all of these mappings produce a new isomorphism.

Note: The smallest possible number is 23 and the largest is 99. The ASCII code for 23 is not a printable character. An inspection of the ASCII table will explain why it was decided to add 23.

The 10,691 color combinations obtained by the algorithm described in Section 5 were listed in order according to their ASCII characters. The first entry, .:Z[yz, has sixty isomorphisms. These were removed from the list. The head of the list then became .:Zppp; its isomorphisms are determined and removed from the list. The process was then continued until the list was exhausted. This process then identified the 148 disjoint sets of isomorphisms.

Shown in Figure 4 are an edge solution and a center solution for the color combination of Figure 1. The solutions are identified by a three-digit number indicating the color of the buried cube and the two "edge" colors. Different solutions often have the same three-digit identification.

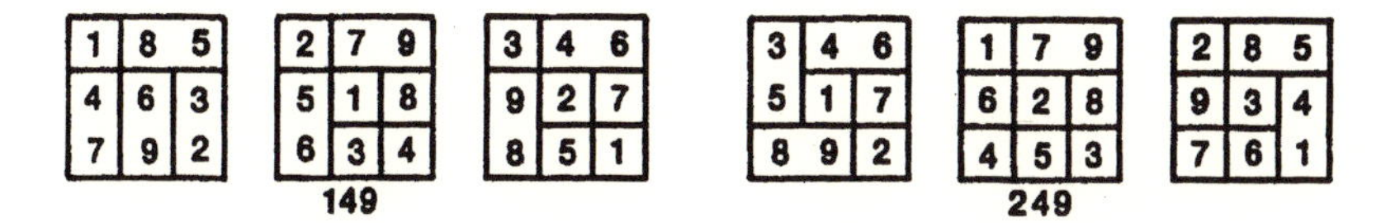

Figure 4. Typical solutions.

If the solutions are to be catalogued, then additional criteria for recording solutions are recommended.

The edge solution is interesting in that new solutions are often obtainable by using two transformations. Whenever the tricube is on an edge and shares a plane with only three dicubes, the plane can be translated to the other side resulting in a new solution. Often two dicubes may be swapped with two others having the same colors. For example, dicubes (4, 7) and (6, 9) in the bottom layer may be swapped with dicubes (4, 6) and (7, 9) on the upper right. Using these transformations, the reader should now be able to determine five more edge solutions.

Stan Isaacs has suggested that graph analysis would be useful in determining the number of solutions. A preliminary investigation shows some merit in utilizing graph analysis to illustrate why some color combinations have no solutions while other color combinations have numerous solutions. Unfortunately, all 148 graphs have not been compared.

Puzzle sets or puzzle solutions may be obtained by contacting the author by E-mail at sivy@ieee.org.

References

[1] Slocum, Jerry, and Botermans, Jack. *Creative Puzzles of the World.* Harry N. Abrams, New York, 1978. 200 pp., hardcover.

[2] Meeus, J., and Torbijn, P. J. *Polycubes.* Distracts 4, CEDIC, Paris, France, 1977. 176 pp., softcover, in French.

Twice: A Sliding Block Puzzle

Edward Hordern

Twice is a new concept in sliding block puzzles: Some blocks are restricted in their movements and can only reach certain parts of the board from particular directions or, in some cases, cannot get there at all.

The puzzle was invented by Dario Uri from Bologna, Italy, and was originally issued in 1989 with the name "Impossible!!" Subsequent additions and improvements made over the next couple of years led to a change of name, to "Twice." The name was chosen because there are two quite different puzzles involving two different blocks numbered 2(a) and 2(b), one being used in each puzzle.

Description

Fixed into the base of the board are four pegs. one in each corner (see Figure 1).

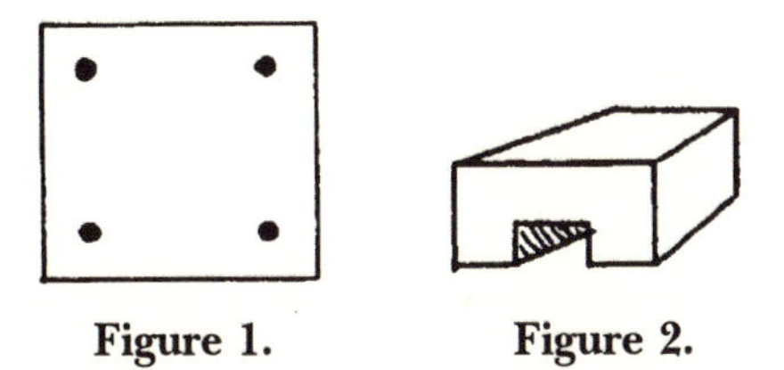

Figure 1. **Figure 2.**

There are nine square blocks (including two numbered 2), but only eight are used in each puzzle. Channels (or grooves) are cut into the bases of some blocks, allowing them to pass over the pegs in the corners (see Figure 2). Blocks can either have a horizontal channel, a vertical channel, both channels, or no channel at all. Blocks A, 7, and 2(b) have a single horizontal channel, and these blocks can only reach the corners from a horizontal direction. Blocks 4 and 2(a) have a single vertical channel and

Written by Edward Hordern and reproduced with the kind permission of Dario Uri.

can only approach the corners from a vertical direction. Blocks 1 and 5 have both channels (in the form of a cross) and can go anywhere. Blocks 3 and 6 have no channels and cannot go into any corner.

The Puzzle

Figure 3 shows the start position. The first puzzle uses block 2(a), and the second puzzle uses block 2(b). The object of the puzzle in each case is to move block A to the bottom right corner. The puzzles are both rated as difficult, the second being the harder of the two. It is quite an achievement just to solve them. For the expert, however, the shortest known solutions are 50 moves for the first puzzle and a staggering 70 moves for the second. Both solutions are believed (but not proved) to be minimum-move solutions.

A		1
2	3	4
5	6	7

Figure 3.

The delight of the first puzzle is that it is very easy to move block A to just above the bottom right corner, only to get hopelessly stuck.... So near, and yet so far! In puzzle 2 it is quite a task to move block A more than a square or two, or to get anywhere at all!

Hints for Solving

In both puzzles the "nuisance" blocks are 3 and 6. Since they can't go into the corners, they must only move in a cross-shaped area. During the solution they must continually be moved "around a corner" to get them out of the way of another block that has to be moved. Once this has been mastered, a plan can be made as to which blocks have to be moved so as to allow block A to pass. The real "problem" blocks are block 7 in the first puzzle and blocks 7 and 2(b) in the second. These blocks, as well as block A, can only move freely up and down the center of the puzzle. This causes something of a traffic jam, which has to be overcome....

Planar Burrs

M. Oskar van Deventer

Usually "burrs" are considered to be three-dimensional puzzles. The most common are the six-piece burrs, which occur in many different designs. The most interesting are those with internal voids, because these can be so constructed that several moves are needed to separate the first piece. Since about 1985 a "Most Moves Competition" for six-piece burrs has been running. Bill Cutler's "Baffling Burr" and Philippe Dubois' "Seven Up" were the first attempts. I am not completely aware of the state of the competition, but as far as I know Bruce Love is the record holder with his "Love's Dozen," which requires twelve moves!

Recently, I received from Tadao Muroi, in Japan, an ingenuously designed puzzle that he called "Four Sticks and a Box." The puzzle has no more than four movable pieces, but nevertheless requires twelve moves (!) to get the first piece out of the box. Muroi wrote me that his idea was inspired by a design of Yun Yananose, who was inspired by "Dead Lock," a puzzle of mine. Though Muroi's puzzle is three-dimensional, all interactions between the four pieces take place in one plane, so in some sense it might be considered a planar burr. However, it cannot be realized as a planar burr because in two dimensions each piece should be disconnected.

True planar burrs are rarely found. The first design I saw is Jeffrey Carter's, depicted in A. K. Dewdney's *Scientific American* column (January 1986, p. 16). Carter's puzzle has four pieces. The puzzle is not very difficult to solve, with only three moves needed to remove the first piece.

The idea of a two-dimensional burr immediately appealed to me, and in April 1986 I made some attempts to find a design of my own. One result is the "Zigzag" planar burr, depicted in Figure 1.

This puzzle has two congruent large pieces and two congruent smaller ones. It takes five moves to separate the first piece. To solve the puzzle, the two large pieces move into each other along a zigzag line, until the two smaller pieces are free. The same movements, in backward order, will separate large pieces.

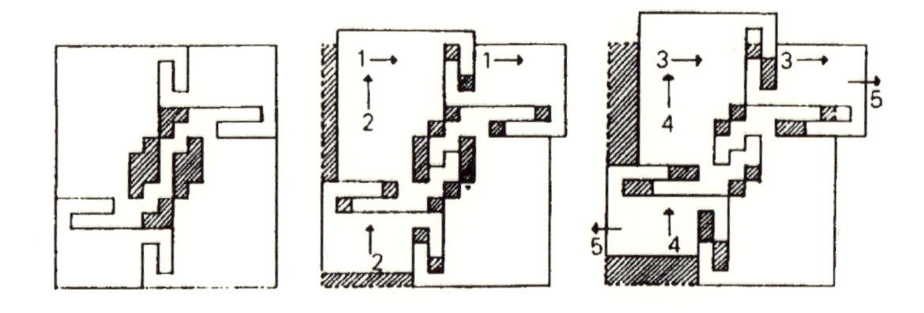

Figure 1. Zigzag.

Some months ago, particularly inspired by Muroi's "Four Sticks" (so closing the circle of mutual inspiration), I took up the challenge again and succeeded in finding a new design of a planar burr, which I have called "Nine and One-Half Moves."

This is is a true two-dimensional burr of only *three* pieces and it needs no less than nine and one-half moves to separate one piece from the other two. The three pieces of the puzzle form a square with internal voids.

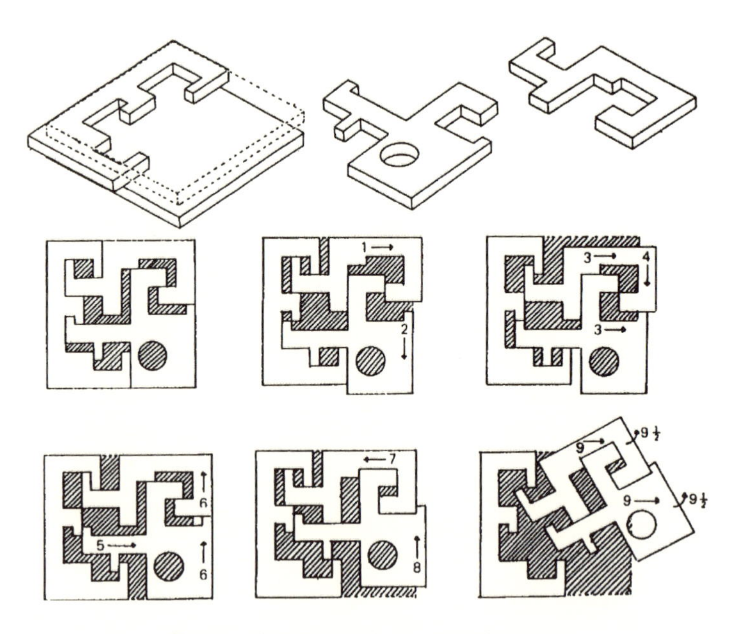

Figure 2. Nine and One-Half Moves.

The pieces, and the moves required to separate the pieces are depicted in Figure 2. The ninth move is a slide-plus-rotate move, so I count it as a move and a half.

In order to prevent us three-dimensional people from cheating, we can glue one piece of the puzzle between two square plates as indicated in the top left corner of Figure 2. By using opaque plates the design is hidden as well. A round hole can be used to hide a coin.

Block-Packing Jambalaya

Bill Cutler

My primary interest over the years has been burr puzzles, but there is another small category of puzzles that is especially intriguing to me. It is 3-dimensional box-packing puzzles where the box and all the pieces are rectangular solids. The number of such puzzles that I am aware of is quite small, but the "tricks," or unique features that the puzzles employ are many and varied. I know of no other small group of puzzles that encompasses such a rich diversity of ideas.

Presented here are 11 such "block-packing" puzzles. The tricks to most of the puzzles are discussed here, but complete solutions are not given.

The puzzles are grouped according to whether there are holes in the assembled puzzle and whether the pieces are all the same or different.

For each puzzle, the total number of pieces is in parentheses. If known, the inventor of the puzzle, date of design, and manufacturer are given.

1. No Holes, All Pieces the Same

"Aren't these puzzles trivial?," you ask. Well, you are not far from being completely correct, but there are some interesting problems. David Klarner gives a thorough discussion of this case in [5]. The following are my favorites:

1. unnamed (44) (Singmaster–Klarner):
   Box: $8 \times 11 \times 21$
   Pieces: (44) $2 \times 3 \times 7$
2. unnamed (45) (de Bruijn):
   Box: $5 \times 6 \times 6$
   Pieces: (45) $1 \times 1 \times 4$

2. No Holes, Limited Number of Piece Types

The puzzles I know of in this category follow a common principal: There are basically two types of pieces – a large supply of one type and a limited supply of another. The pieces of the second type are smaller and easier to use, but must be used efficiently to solve the puzzle. The solver must determine exactly where the second set of pieces must be placed, and then the rest is easy.

3. unnamed (9) (Slothouber–Graatsma):
   Box: $3 \times 3 \times 3$
   Pieces: (3) $1 \times 1 \times 1$, (6) $1 \times 2 \times 2$
4. unnamed (18) (John Conway):
   Box: $5 \times 5 \times 5$
   Pieces: (3) $1 \times 1 \times 3$, (1) $1 \times 2 \times 2$, (1) $2 \times 2 \times 2$, (13) $1 \times 2 \times 4$

In the first of these, the three individual cubes are obviously easy to place, but they must not be wasted. By analyzing "checkerboard" colorings of the layers in the box, it is easy to see that the cubes must be placed on a main diagonal. In the second design, the three $1 \times 1 \times 3$ pieces must be used sparingly. The rest of the pieces, although not exactly alike, function similarly to the $1 \times 2 \times 2$ pieces in the first puzzle. See [2] or [5] for more information.

3. No Holes, Pieces Mostly Different

5. Quadron (18) (Jost Hanny, Naef):
   Box 1: $5 \times 7 \times 8$
   Pieces: $2 \times 3 \times 3$, $2 \times 3 \times 5$, $2 \times 4 \times 5$, $2 \times 4 \times 6$, $3 \times 3 \times 4$, $3 \times 3 \times 5$, $3 \times 3 \times 7$
   Box 2: $5 \times 7 \times 10$
   Pieces: $1 \times 3 \times 4$, $1 \times 3 \times 6$, $1 \times 3 \times 7$, $1 \times 3 \times 10$, $1 \times 4 \times 5$, $2 \times 3 \times 4$, $2 \times 3 \times 6$, $2 \times 3 \times 7$, $2 \times 4 \times 7$, $3 \times 3 \times 3$, $4 \times 4 \times 4$
   Box 3: $7 \times 9 \times 10$
   Pieces: all 18 pieces from first two boxes

Quadron does not use any special tricks that I am aware of, but it does make a nice set of puzzles. The 18 pieces are all different, and the three boxes offer a wide range of difficulty. The small box is very easy. The seven pieces can be placed in the box in ten different ways, not counting rotations and/or reflections. A complete, rigorous analysis of the puzzle can be done by hand in about 15 minutes. The middle-size box is difficult –

there is only one solution. The large box is moderately difficult, and has many solutions.

Quadron also makes for a nice entrance into the realm of computer analysis of puzzles and the limitations of such programs. The programmer can use algorithms that are used for pentominoe problems, but there are more efficient algorithms that can be used for block-packing puzzles. I wrote such a program on my first computer, a Commodore 64. The program displayed the status of the box at any instant using color graphics. I painted pieces of an actual model to match the display. The result was a fascinating demonstration of how a computer can be used to solve such a puzzle. The Commodore 64 is such a wonderously slow machine – when running the program in interpreter BASIC, about once a second a piece is added or removed from the box! Using compiled BASIC, the rate increases to 40 pieces/second.

When running these programs on more powerful computers, the difference between the three boxes is stunning: The first box can be completely analyzed in a small fraction of a second. The second box was analyzed in about a minute of mainframe computer time. In early 1996, I did a complete analysis of the third box. There are 3,450,480 solutions, not counting rotations and reflections. The analysis was done on about 20 powerful IBM workstations. The total CPU time used was about 8500 hours, or the equivalent of one year on one machine. By the end of the runs, the machines had constructed 2 1/2 trillion different partially filled boxes.

6. Parcel Post Puzzle (18) (designer unknown; copied from a model in the collection of Abel Garcia):

   Box: $6 \times 18 \times 28$

   Pieces: all pieces are of thickness 2 units; the widths and lengths are

   4×9, 5×18, 5×21, 6×7, 6×10, 6×13, 7×18, 8×18, 9×11, 9×13, 10×11, 11×11 and two each of 5×9, 7×8, and 7×13.

Since all the pieces are of the same thickness and the box depth equals three thicknesses, it is tempting to solve the puzzle by constructing three layers of pieces. One or two individual layers can be constructed, but the process cannot be completed. The solution involves use of the following obvious trick (is that an oxymoron?): Some piece(s) are placed sideways in the box. Of the 18 pieces, 10 are too wide to fit into the box sideways and 4 are of width 5, which is no good for this purpose. This leaves 4 pieces that might be placed sideways. There are four solutions to the puzzle, all very similar, and they all have three of these four pieces placed sideways.

7. Boxed Box (23) (Cutler, 1978, Bill Cutler Puzzles):

   Box: $147 \times 157 \times 175$

Pieces: $13 \times 112 \times 141$, $14 \times 70 \times 75$, $15 \times 44 \times 50$, $16 \times 74 \times 140$, $17 \times 24 \times 67$, $18 \times 72 \times 82$, $19 \times 53 \times 86$, $20 \times 40 \times 92$, $21 \times 52 \times 65$, $22 \times 107 \times 131$, $23 \times 41 \times 73$, $26 \times 49 \times 56$, $27 \times 36 \times 48$, $28 \times 55 \times 123$, $30 \times 54 \times 134$, $31 \times 69 \times 78$, $33 \times 46 \times 60$, $34 \times 110 \times 135$, $35 \times 62 \times 127$, $37 \times 83 \times 121$, $38 \times 42 \times 90$, $45 \times 68 \times 85$, $57 \times 87 \times 97$

The dimensions of the pieces are all different numbers. The pieces fit into the box with no extra space. The smallest number for which this can be done is 23. There are many other 23-piece solutions that are combinatorially different from the above design. Almost 15 years later, this puzzle still fascinates me. See [1] or [3] for more information.

4. Holes, Pieces the Same or Similar

8. Hoffman's Blocks (27) (Dean Hoffman, 1976)
   Box: $15 \times 15 \times 15$
   Pieces: (27) $4 \times 5 \times 6$

This sounds like a simple puzzle, but it is not. The extra space makes available a whole new realm of possibilities. There are 21 solutions, none having any symmetry or pattern. The dimensions of the pieces can be modified. They can be any three different numbers, where the smallest is greater than one-quarter of the sum. The box is a cube with side equal to the sum. I like the dimensions above because it tempts the solver to stack the pieces three deep in the middle dimension. See [4].

9. Hoffman Junior (8) (NOB Yoshigahara, 1986, Hikimi Puzzland)
   Box: $19 \times 19 \times 19$
   Pieces: Two each of $8 \times 9 \times 10$, $8 \times 9 \times 11$, $8 \times 10 \times 11$, $9 \times 10 \times 11$

5. Holes, Pieces Different

10. Cutler's Dilemma, Simplified (15) (Cutler, 1981, Bill Cutler Puzzles)
    Box: $40 \times 42 \times 42$
    Pieces: $9 \times 19 \times 26$, $9 \times 20 \times 20$, $10 \times 11 \times 42$, $10 \times 12 \times 26$, $10 \times 16 \times 31$, $10 \times 19 \times 25$, $10 \times 19 \times 26$, $11 \times 11 \times 25$, $11 \times 12 \times 42$, $11 \times 16 \times 19$, $11 \times 17 \times 30$, $11 \times 19 \times 25$, $12 \times 17 \times 19$, $16 \times 19 \times 21$, $17 \times 19 \times 21$

The original design of Cutler's Dilemma had 23 pieces and was constructed from the above, basic, version by cutting some of the pieces into two or three smaller pieces. The net result was a puzzle that is extremely difficult. I will not say anything more about this design except that the trick involved is different from any of those used by the other designs in this paper.

6. Miscellaneous

11. Melting Block (8-9) (Tom O'Beirne)
    Box: $58 \times 88 \times 133$
    Pieces: $19 \times 29 \times 44$, $19 \times 29 \times 88$, $19 \times 58 \times 44$, $38 \times 29 \times 44$, $19 \times 58 \times 88$, $38 \times 29 \times 88$, $38 \times 58 \times 44$, $38 \times 58 \times 88$
    plus a second copy of $19 \times 29 \times 44$

The Melting Block is more of a paradox than a puzzle. The eight pieces fit together easily to form a rectangular block $57 \times 87 \times 132$. This fits into the box with a little room all around, but seems to the casual observer to fill up the box completely. When the ninth piece is added to the group, the pieces can be rearranged to make a $58 \times 88 \times 133$ rectangular solid. (This second construction is a little more difficult.) This is a great puzzle to show to "non-puzzle people" and is one of my favorites.

By the way, one of the puzzles listed above is impossible. I won't say which one (it should be easy to figure out). It is a valuable weapon in every puzzle collector's arsenal. Pack all the pieces, except one, into the box, being sure that the unfilled space is concealed at the bottom and is stable. Place the box on your puzzle shelf with the remaining piece hidden behind the box. You are now prepared for your next encounter with a boring puzzle-nut. (No, readers, this is not another oxymoron, but rather a tautology to the 99% of the world that would never even have started to read this article.) Pick up the box and last piece with both hands, being careful to keep the renegade piece hidden from view. Show off the solved box to your victim, and then dump the pieces onto the floor, including the one in your hand. This should keep him busy for quite some time!

References

[1] W. Cutler, "Subdividing a Box into Completely Incongruent Boxes," *J. Recreational Math.,* 12(2), 1979–80, pp. 104–111.

[2] M. Gardner, Mathematical Games column in *Scientific American,* February 1976, pp. 122–127.

[3] M. Gardner, Mathematical Games column in *Scientific American,* February 1979, pp. 20–23.

[4] D. Hoffman, "Packing Problems and Inequalities," in *The Mathematical Gardner,* edited by D. Klarner (Wadsworth International, 1981), pp. 212–225.

[5] D. Klarner, "Brick-Packing Puzzles," *J. Recreational Math.*, 6(2), Spring 1973, pp. 112–117.

Written on the occasion of the Puzzle Exhibition at the Atlanta International Museum of Art and Design and dedicated to Martin Gardner.

Classification of Mechanical Puzzles and Physical Objects Related to Puzzles

James Dalgety and Edward Hordern

Background. "Mechanical Puzzles" is the descriptive term used for what are also known as "Chinese Puzzles." Several attempts have been made to classify mechanical puzzles, but most attempts so far have either been far too specialized in application or too general to provide the basis for a definitive classification. Many people have provided a great deal of help, but particular thanks are due to Stanley Isaacs, David Singmaster, and Jerry Slocum.

Objective. To provide a logical and easy-to-use classification to enable non-experts to find single and related puzzles in a large collection of objects, and patents, books, etc., related to such objects. (As presented here, while examples are given for most groups, some knowledge of the subject is required.)

Definitions. A *puzzle* is a problem having one or more specific objectives, contrived for the principle purpose of exercising one's ingenuity and/or patience. A *mechanical puzzle* is a physical object comprising one or more parts that fall within the above definition.

Method. A puzzle should be classified by the problem that its designer intended the solver to encounter while attempting to solve it. Consider a three-dimensional (3-D) interlocking assembly in the form of a cage with a ball in the center. The fact that the instructions request the would-be solver to "remove the ball" does not change the 3-D assembly into an opening puzzle. The disassembly and/or reassembly of the cage remains the primary function of the puzzle. An interlocking puzzle should be classified according to its interior construction, rather than its outward appearance (e.g., a wooden cube, sphere, barrel, or teddy bear may all have similar Cartesian internal construction and so should all be classified as Interlocking–Cartesian).

For updated information and illustrations, go to http://puzzlemuseum.com.

In cases where it seems possible to place a puzzle in more than one category, it must be classified in whichever is the most significant category. A few puzzles may have to be cross-referenced if it is absolutely necessary; usually, however, one category will be dominant.

A good example of multiple-class puzzles is the "Mazy Ball Game" made in Taiwan in the 1990s. It is based on a 3×3 sliding block puzzle under a clear plastic top. The pieces have L-shaped grooves, and a ball must be rolled up a ramp in the lower right onto one of the blocks – the ball must be moved from block to block, and the blocks themselves must be slid around so that the ball can exit at the top left. Thus the puzzle requires Dexterity, Sequential movement, and Route-finding. It would be classified as Route-finding because, if the route has been found, then the dexterity and sequential movement must also have been achieved.

A puzzle will be referred to as two-dimensional (2-D) if its third dimension is irrelevant (e.g., thickness of paper or plywood or an operation involving a third dimension such as folding). Most standard jigsaws are 2-D, although jigsaws with sloping cuts in fact have a relevant third dimension, so they must be classed as 3-D. It will be noted that the definition of a puzzle excludes the infant's "posting box," which, while perhaps puzzling the infant, was contrived only to educate and amuse; it also excludes the archer attempting to get a bull's-eye, the exercise of whose ingenuity is entirely incidental to the original warlike intent of the sport. Also excluded are puzzles that only require paper and pencil (e.g., crossword puzzles), unless they are on or part of some physical object.

It is understood that specialist collectors will further subdivide the subclasses to suit their own specialized needs. For example, Tanglement-Rigid & Semi-Rigid is awaiting a thorough study of the topology of wire puzzles.

The full abbreviations consist of three characters, hyphen, plus up to four characters, such as "INT-BOX." These are the standard abbreviations for the classes that have been chosen for relative ease of memory and conformity with most computer databases.

The fourteen main classes are as follows:

- *Dexterity Puzzles (DEX)* require the use of manual or other physical skills in their solution.
- *Routefinding Puzzles (RTF)* require the solver to find either any path or a specific path as defined by certain rules.
- *Tanglement Puzzles (TNG)* have parts that must be linked or unlinked. The linked parts, which may be flexible, have significant

freedom of movement in relation to each other, unlike the parts of an interlocking puzzle.

- *Opening Puzzles (OPN)* are puzzles in which the principle object is to open it, close it, undo it, remove something from it, or otherwise get it to work. They usually comprise a single object or associated parts such as a box with its lid, a padlock and its hasp, or a nut and bolt. The mechanism of the puzzle is not usually apparent, nor do they involve general assembly/disassembly of parts that interlock in 3-D.
- *Interlocking Puzzles (INT)* interlock in three dimensions; i.e., one or more pieces hold the rest together, or the pieces are mutually self-sustaining. Many clip-together puzzles are "non-interlocking."
- *Assembly Puzzles (Non-Interlocking) (ASS)* require the arrangement of separate pieces to make specific shapes without regard to the sequence of that placing. They may clip together but do not interlock in 3-D. Some have a container and are posed as packing problems.
- *Jigsaw Puzzles (JIG)* are made from cut or stamped-out pieces from a single complete object, and the principle objective is to restore them to their unique original form.
- *Pattern Puzzles (PAT)* require the placing or arrangement of separate pieces of a similar nature to complete patterns according to defined rules. The pattern required may be the matching of edges of squares, faces of a cube, etc. The pattern may be color, texture, magnetic poles, shape, etc. Where the pattern is due to differences in shape, the differences must be sufficiently minor so as not to obscure the similarity of the pieces.
- *Sequential Movement Puzzles (SEQ)* can be solved only by moves that can be seen to be dependent on previously made moves.
- *Folding and Hinged Puzzles (FOL)* have parts that are joined together and usually do not come apart. They are solved by hinging, flexing, or folding.
- *Jugs and Vessels (JUG).* Vessels having a mechanical puzzle or trick in their construction that affects the filling, pouring, or drinking therefrom.
- *Other Types of Mechanical Puzzles and Objects (OTH).* This group is for puzzle objects that do not easily fall into the above categories and cannot be categorized into sufficiently large groups to warrant their own major class. Included in this group are Balancing, Measuring, Cutting, Math, Logic, Trick, Mystery, and Theoretical Puzzles. Also, provision is made for puzzles pending classification.
- *Ambiguous Pictures and Puzzling Objects (AMB).* Puzzles in which something appears impossible or ambiguous.

- *Ephemera (EPH).* This category has been included because most puzzle collections include related ephemera, which, while not strictly puzzles, need to be classified as part of the collection.

A detailed classification with second-level classes is given in the separate table. Puzzle Class Abbreviations (PZCODE) are standardized to a maximum of eight characters: XXX-YYYY, where XXX is the main class and YYYY is the sub-class. Examples of puzzles in each class are given in the right-hand column.

Proposed Developments

Prior to 1998 the subclasses attempted to incorporate the number of dimensions, the type of structure, and any group/non-group moves. This has resulted in an unwieldy list. We are now working on defining what should be included automatically in the subclasses where they are relevant. We hope to reduce the number of sub-classes substantially in the near future.

Number of Dimensions. 2-D, 3-D, 2-D on 3-D, 2-D to 3-D, 2-D, 3-D, and 4-D. Required for INT JIG ASS PAT RTF SEQ FOL.

Group Moves. Whether needed, not needed, or partly used in solution. Required for SEQ.

Number of Pieces. Required for INT JIG ASS PAT.

Type of Piece Structure. Required for INT ASS PAT.

Examples of approved structure adjectives that may be used include

- *Identical:* All pieces identical
- *Cartesian:* Three mutually perpendicular axes
- *Diagonal:* Pieces rotated 45 degrees along their axis
- *Skewed:* Like a squashed puzzle
- *Polyhedral*
- *Geometrical:* Non-Cartesian but geometrical structures
- *Organic:* Amorphously shaped pieces
- *Ball:* Pieces made from joined spheres
- *Rod:* Square, hexagonal, triangular, etc.
- *Linked:* Pieces joined together by hinges, strings, ribbons, etc.

Thus a standard six-piece burr uses square rods in a Cartesian structure, and the standard Stellated Rhombic Dodecahedron has a six-piece Diagonal Cartesian structure.

Guide to Making a Catalogue or Database for Puzzle Collections

Headings for cataloguing puzzle collections could include the following items. In practice, without paid curators, it is probably advisable to be selective and limit the amount of information recorded.

Generalized Information

- Generic name of puzzle or objective if not obvious. **required
- Class + Subclass **required

Information Specific to This Object

- Theme or advertisement (subject)
- Materials
- Dimensions A, B, C; d = diameter. **A required for scale
- What the dimension refers to, i.e., box, envelope, biggest piece, assembled puzzle
- If powered, i.e., battery, clockwork, electric, solar
- Patent and markings
- Notes, references, designer
- Manufacturer or publisher's name and country
- Type of manufacturing, i.e., mass produced (over 5000 pieces made), craft made (commercial but small volume), homemade or tribal
- Year of manufacture **required
- Country of manufacture if different from publisher's country
- Manufacturer's series name
- Manufacturer's product name
- Number in complete set, if known
- Number of set in collection
- A picture or photograph

Information Specific to This Collection

- Location/cabinet/bin
- Acquisition
- Number
- Condition (Excellent, Good, Fair, Poor)
- Condition qualifying note, i.e., puzzle may have a cracked glass top or be missing one piece, but otherwise be in excellent condition.
- Source
- Date
- Cost
- Insurance value

Detailed Puzzle Classification

Class	Description	Example
DEX-	**Dexterity Puzzles**	
DEX-UNCA	Dexterity or other physical skills in their solution	Cup and Ball, "Le Pendu," "Theo der Turnier," Tomy's "Crazy Maze," puzzles using tops
DEX-BALL	Dexterity; plain balls into holes	Pentangle "Roly-Poly" puzzles
DEX-SDRY	Dexterity with sundry objects and/or obstacles	Ramps, bridges, jumping beans, etc.
DEX-LQOB	Liquid objects	Mercury manipulation
DEX-INLQ	Dexterity in liquid	Water-filled puzzles
DEX-MIRR	Indirect viewing	View by mirror
DEX-MECH	Mechanized	Tomy's "Pocketeers"
DEX-TOOL	Using tools and magnetic tools	
DEX-RTFL	Route following dexterity	
DEX-HIDD	Objects concealed from view	Four Generations "Ball in Block," Engel's "Black Box"
DEX-ELEC	Electrica and electronic dexterities	
DEX-PINB	Pinball-related dexterities	Bagatelle
DEX-OTHR	Other dexterities, pneumatic operation	
RTF-	**Route-Finding Puzzles**	
RTF-CHNG	Route-finding with changing path and/or Complex Traveller	"Frying Pan," "Yankee," "Tandem Maze" (complex), "Bootlegger" (complex)
RTF-STEP	Route-finding step mazes	Ring and hole mazes, "Pike's Peak"
RTF-UNIC	Unicursal route-finding	Icosian Game, Königsburg Bridges
RTF-SHOR	Shortest route	
RTF-CPLX	Complex route mazes with special objectives	"Worried Woodworm," colour mazes, number totalling mazes, avoiding objects, visiting places en route.
RTF-2D	Route mazes 2-D (any path)	Most hedge mazes

RTF-2D3D	Route mazes 2-D on 3-D surfaces (any path)	Maze on surface of cube
RTF-3D	Route mazes 3-D (any path)	Some hedge mazes, ball in 4× 4 cube of cubelets
TNG-	**Tanglement Puzzles**	
TNG-RIGI	Tanglement of rigid and semi-rigid parts	Wire puzzles, cast "ABC," Chinese rings, puzzle rings
TNG-R&F	Tanglement of rigid and flexible parts	Hess wire puzzles, Dalgety's "Devil's Halo"
TNG-FLEX	All flexible	Leather tanglement puzzles
OPN-	**Opening Puzzles**	
OPN-BOX	Opening containers	Boxes, purses
OPN-LOCK	Opening locks	Padlocks
OPN-HID	Opening/finding hidden compartments not originally designed as puzzles	Chippendale tea chests, poison rings
OPN-OTHR	Opening other objects	Nuts and bolts, knives, pens, cutlery, Oskar's keys, Oscar's "Dovetail," "Hazelgrove Box"
INT-	**Interlocking Puzzles**	
INT-BOX	Boxes that disassemble	Strijbos aluminium burr box
INT-CART	Cartesian (has parts along three mutually perpendicular axes)	Burrs, "Mayer's Cube," "Margot Cube," Cutler's burr in a glass
INT-POLY	Interlocking polyhedral	Coffin's "Saturn"
INT-SHAP	Other rigid shapes	JWIP and keychain animals, berrocals, "Tak-it-Apart," "Nine of Swords," keychain cars
INT-TENS	Tensegrity structures in which compression and tension elements separate	"Plato's Plight"
ASS-	**Assembly Puzzles**	
ASS-MAT	Match stick puzzles and tricks	
ASS-2D	2-D Assembly	Tangram, Pentominoes, gears, checkerboards, "T" puzzle
ASS-CART	3-D Cartesian Assemblies	Polycubes, Soma, "Hoffman" cube, O'Beirne's "Melting Block"

ASS-POLY	3-D Assembly Polyhedra and Spheres	Ball pyramids, Squashed Soma, nine-piece ivory cube
ASS-SHAP	3D Assembly Other Shapes	Pack the Plums, Apple and Worms, "Managon," "Even Steven," "Phoney Baloney," Gannt's "Chandy"
JIG-	**Jigsaw Puzzles**	
JIG-STD	Standard jigsaws	Can include double-sided puzzles
JIG-3D	3-D jigsaws	
JIG-2DID	2-D jigsaws with identical pieces for secondary objectives	"Shmuzzles" tesselations
JIG-PART	2-D jigsaws which only partially cover the plane	Bilhourd's jigsaws
JIG-SLOP	2-D jigsaws with non-perpendicular/sloping cuts	
JIG-LAYR	Multiple-layer 2-D jigsaws	"Sculpture Puzzles"
JIG-2D3D	2-D jigsaws with parts that can be made into 3-D objects	"Toyznet"
JIG-BLOX	Picture cubes/blocks	
PAT-	**Pattern Puzzles**	
PAT-2DPG	Pattern arrangements of points, pegs, or pieces, according to predetermined rules	Queens on chess board, BlackBox, Josephus, Waddington's "Black Box"
PAT-STIX	Patterns of sticks	Match puzzles, Jensen's "Tricky Laberint"
PAT-NUM	Pattern: arrangements of number	Magic squares, number puzzles
PAT-2DEG	2-D matching edge patterns	Heads and tails
PAT-2D	Arrangement of 2-D pieces on 2-D surface to make a pattern according to predetermined rules	"Testa"
PAT-STAK	Stacking, overlapping, and weaving 2-D patterns	Stacking transparent layers, "Lapin," Loyd's Donkeys, weaving puzzles
PAT-2D3D	Patterns with 2-D parts on 3-D surface	"Dodeca"

PAT-3D	3-D pattern puzzles with separate parts	"Instant Insanity," Waddington's "Kolor Kraze," Skor Mor "Instant Indecision," Chinese balls in ball, Oscar's "Solar System," Laker Cube
PAT-LINK	Linked part pattern puzzles	Bognar's planets, Panel Nine, "Dodecahedru" – rotating faces of Dodecahedron
SEQ-	**Sequential Puzzles**	
SEQ-PLAC	Sequential placement	Psychic Pz, Fit Pz
SEQ-RIVR	Sequential river crossing	"Wolf, Sheep and Cabbage"
SEQ-HOPP	Sequential hopping and jumping	Solitaire, Tower of Hanoi, counter and peg moving puzzles
SEQ-SL2D	Sequential sliding and shunting in 2-D	15s Pz., Tit-Bit's "Teasers"
SEQ-SL3D	Sliding and shunting in 3-D (only single piece moves needed)	"Inversions"
SEQ-SLRO	Sliding and shunting with mechanical or rotating parts (single piece moves and group moves needed)	"Tower of Babel," "Missing Link," "Backspin," "Turntable Train", Tomy's "Great Gears"
SEQ-RT2D	Sequential rotating in 2-D (group moves only)	Raba's "Rotascope," Rubik's "Clock"
SEQ-RT3D	Sequential rotating and/or mechanical in 3-D (group moves only)	Rubik's Cube, "Masterball," "Jugo Flower," "Orbit," "Kaos"
SEQ-ROLL	Sequential rolling	Rolling eight cubes
SEQ-MMEC	Sequential miscellaneous mechanical	"The Brain," "Hexadecimal," "Spin Out"
FOL-	**Folding Puzzles**	
FOL-SING	Folding single-part puzzles	"Why Knots," Möbius strips
FOL-HGOP	Folding hinged parts in open loop	Rubik's "Snake," strung cubes
FOL-HGCL	Hinged parts in closed loop	Flexagons, Rubik's "Magic," "Flexicube"
FOL-HSEP	Folding hinged parts that separate	"Clinch Cube"

FOL-SH2D	Folding sheets and strips into 2-D solution	Map folding, "Jail Nixon"
FOL-SH3D	Folding sheets and strips into 3-D shapes	Strip polyhedra
JUG-	**Puzzle Jugs**	
JUG-STD	Puzzle mugs standard	Standard "block holes and suck" solution
JUG-CPLX	Complex mugs requiring special manipulation	
JUG-BASE	Pour from base	"Jolly Jugs"
JUG-NLID	Lidless wine jugs	Cadogan teapots, Chinese winepots
JUG-OTHR	Self-pouring and other patents	Royale's patent
OTH-	**Other Types of Puzzle**	
OTH-ELEC	Electrical and electronic (non-dexterous)	Luminations
OTH-BAL	Balancing (non-dexterity)	"Columbus Egg"
OTH-MEAS	Measuring and weighing puzzles	Jugs and liquids, 12 Golf Balls, Archimedes gold
OTH-CUT	Cutting puzzles	Cork for three holes, five square puzzle
OTH-RIDD	Riddles	19th century riddle prints
OTH-WORD	Word puzzles	Anagrams, rebus plates and prints, crosswords on bathroom tissue
OTH-MATH	Mathematical puzzles (excluding number pattern arrangements)	
OTH-LOGI	Logic puzzles	Cartoon pictures to arrange in order
OTH-TRIK	Trick or catch puzzles (solution needs subterfuge)	"Infernal Bottle"
OTH-MAGI	Conjuring tricks presented as puzzles	Disappearing coin slide box
OTH-MYST	Objects whose function or material is a mystery	Creteco spacers
OTH-SET	Sets of puzzles of mixed type	
OTH-THEO	Puzzles whose existence is only theoretically possible . . .	such as 4-D puzzles, or those that can only be represented on a computer

OTH-PEND	Pending Classification !!!	Puzzles awaiting classification
AMB-	**Ambiguous**	
AMB-POBJ	Paradoxical objects (objects that apparently cannot be made)	Arrow through bottle, impossible dovetails, Oskar's "Escher Puzzle," Penrose triangle
AMB-VANI	Vanishing images	"Vanishing Leprechauns," Hooper's paradox
AMB-DIST	Distortions	Anamorphic pictures
AMB-ARCH	Archimboldesque objects	Pictures and objects of one subject made from completely unrelated objects
AMB-HIDD	Hidden image pictures (no manipulation required)	Devinettes (obscure outlines), "Spot the Difference," random dot stereograms
AMB-HMAN	Hidden image pictures (manipulation required)	"Naughty Butterflies," "Find the 5th Pig" needing coloured overlays, soot on unglazed part of ashtray
AMB-TURN	Pictures that require turning to show different images	Landscape turned to make a portrait, Topsey Turveys, OHOs, courtship/matrimony
AMB-ILLU	Perception illusions	Optical illusions, weight illusions
EPH-	**Ephemera related to puzzles**	
EPH-WIWO	Wiggle Woggle HTL	Hold to light cards producing movements by shadows
EPH-MICR	Micro Printing	Images concealed by extreme smallness
EPH-MOIR	Moiré effect	Puzzles/effects produced by moiré/fringe effects
EPH-HTL	Hold to light	Protean views, hold to light advertisements
EPH-HEIR	Heiroglyphs (non-rebus)	Obscure non-rebus heiroglyphic prints
EPH-ANAG	Anaglyphs	Requiring red and green glasses for either 3-D or movement effects
EPH-STRP	Strip prints (three views in one frame)	Framed strip prints showing different views from different directions

EPH-OTHR	Other puzzle-related ephemera	
XXX-XXX	Lost records	For database integrity only
DEL	Deleted record	Puzzle disposed of; no longer in collection

Part III

Mathemagics

A Curious Paradox

Raymond Smullyan

Consider two positive integers x and y, one of which is twice as great as the other. We are not told whether it is x or y that is the greater of the two. I will now prove the following two obviously incompatible propositions.

Proposition 1. *The excess of x over y, if x is greater than y, is greater than the excess of y over x, if y is greater than x.*

Proposition 2. *The two amounts are really the same (i.e., the excess of x over y, if x is greater than y, is equal to the excess of y over x, if y is greater than x).*

Proof of Proposition 1. Suppose x is greater than y. Then $x = 2y$, hence the excess of x over y is then y. Thus the excess of x over y, if x is greater than y, is y. Now, suppose y is greater than x. Then $x = \frac{1}{2}y$, hence the excess of y over x is then $y - \frac{1}{2}y = \frac{1}{2}y$. Thus the excess of y over x, if y is greater than x, is $\frac{1}{2}y$. Since y is greater than $\frac{1}{2}y$, this proves that the excess of x over y, if x is greater than y, is greater than the excess of y over x, if y is greater than x. Thus Proposition 1 is established. □

Proof of Proposition 2. Let d be the difference between x and y — or what is the same thing, the lesser of the two. Then obviously the excess of x over y, if x is greater than y, is d, and the excess of y over x, if y is greater than x, is again d. Since $d = d$, Proposition 2 is established! □

Now, Propositions 1 and 2 can't both be true! Which of the two propositions do you actually believe?

Most people seem to opt for Proposition 2. But look, suppose y, say, is 100. Then the excess of x over y, if x is greater than y, is certainly 100, and the excess of y over x, if y is greater than x, is certainly 50 (since x is then 50). And isn't 100 surely greater than 50?

A Powerful Procedure for Proving Practical Propositions

Solomon W. Golomb

The eminent Oxford don, Charles Lutwidge Dodgson, demonstrated the applicability of formal mathematical reasoning to real-life situations with such incontrovertible rigor as evidenced in his syllogism:

All Scotsmen are canny.
All dragons are uncanny.
Therefore, no Scotsmen are dragons.

While his logic is impeccable, the conclusion (that no Scotsmen are dragons) is not particularly surprising, nor does it shed much light on situations that we are likely to encounter on a daily basis.

The purpose of this note is to exploit this powerful proof methodology, introduced by Dodgson, to a broader range of human experience, with special emphasis on obtaining conclusions having political or moral significance.

Theorem 1. *Apathetic people are not human beings.*

Proof: All human beings are different.
All apathetic people are indifferent.
Therefore, no apathetic people are human beings.

Theorem 2. *All incomplete investigations are biased.*

Proof: Every incomplete investigation is a partial investigation.
Every unbiased investigation is an impartial investigation.
Therefore, no incomplete investigation is unbiased.

Numerous additional examples, closely following Dodgson's original model, could be adduced. However, our next objective is to broaden the approach to encompass other models of mathematical proof. For example, it is a well-established principle that a property P is true for *all* members of

Reprinted from *Mathematics Magazine*, Vol. 67, No. 5, December 1994, p. 383.

a set *S* if it can be shown to be true for an *arbitrary* member of the set *S*. We exploit this to obtain the following important result.

Theorem 3. *All governments are unjust.*

Proof: To prove the assertion for *all* governments, it is sufficient to prove it for an *arbitrary* government. If a government is arbitrary, it is obviously unjust. And since this is true for an arbitrary government, it is true for all governments.

Finding further theorems of this type, and extending the method to other models of mathematical proof, is left as an exercise for the reader.

Misfiring Tasks

Ken Knowlton

Years ago I went on a collecting trip, visiting non-innumeretic friends asking for contributions of a very particular kind. Martin Gardner gave me some help, not as much as I expected; others scraped together a token from here and there. I browsed, I pilfered, but the bottom of my basket ended up barely covered and, except for the accumulated webs and dust, so it remains. At this point, before throwing it out, I pass it around once more for contributions.

When a cold engine starts, it often misfires a few times before running smoothly. There are mathematical tasks like that: Parameterized in n, each can be accomplished for some hit-and-miss pattern of integer values until, after some last "misfiring" value, things run smoothly, meaning simply that the task can be performed for all higher integers. One such task is to partition a square into n subsquares (i.e., using all of its area, divide it into n non-overlapping subsquares). This can in fact be performed for $n = 4$, 6, 7, 8, ... where 5 is "missing" from the sequence.

A devilishly difficult kind of mathematical problem, I thought, might be to pose such a question in reverse, e.g., What is a task parameterized in n that last misfires for $n = 6$?, or for $n = 9$?, etc. More precisely, a task that last misfires for N

- can be done for at least one earlier n ($1 < n < N$);
- cannot be done for $n = N$;
- can be done for all $n > N$;

and, to be sporting,

- the task must not be designed in an ad hoc way with the desired answer built in – must not, for example, involve an equation with poles or zeros in just the right places;
- it must be positively stated, *not* for example "prove the impossibility of ...";
- it must at least seem to have originated innocently, not from a generative construct such as "divide n things into groups of 17 and 5."

For a mathematician or logician, these are attrociously imprecise specifications. But here, for me at least, lies the intriguing question: Will we agree that one or another task statement lies within the spirit of the game? This is, of course, a sociological rather than mathematical question; my guess is that the matter lies somewhere on the non-crispness scale between agreement on "proof" and agreement on "elegance."

Meager as the current collection is, its members do exhibit a special charm. I have answers only for $N = 5, 6, 7, 9, 47$, and 77. Some of these are well known, others quite esoteric:

(5) Stated above, well known, with a rather obvious proof: Partition a square into n subsquares.

(6) Divide a rectangle into n disjoint subrectangles without creating a composite rectangle except for the whole (Frank Sinden).

(7) Design a polyhedron with n edges. (Generally known. The tetrahedron has 6 edges, the next have 8, 9, ... edges.)

(9) Cut a square into n acute triangles with clean topology: no triangle's vertex may lie along the side of another triangle. Possible for 8, 10, 11, 12 .... (Charles Cassidy and Graham Lord, even after developing a complete proof, ask themselves "Why is 9 missing?")

(9) Place n counters on an infinite chessboard such that each pair exhibits different numbers of row-mates and/or different numbers of column-mates. (Impossible for n = 2, 5, or 9; conjectured by Ken Knowlton, proved by Ron Graham. One of Martin Gardner's books contains essentially this problem as a wire-identification task, but the stated answer implicitly but erroneously suggests that the task can be performed for any n).

(47) Cut a cube into n subcubes (called the "Hadwiger" problem, tracked for years by Martin Gardner, finally clinched independently by Doris Rychener and A. Zbinder who demonstrated a dissection into 54 subcubes).

(77) Partition n into distinct positive integers whose reciprocals sum to one, e.g., such a partition for 11 is 2, 3, 6 since $2 + 3 + 6 = 11$ and $\frac{1}{2} + \frac{1}{3} + \frac{1}{6} = 1$. (Ron Graham is the only person I know who would think of such a problem, and who did, and who then went on to prove that 77 is the largest integer that cannot be so partitioned.)

What we have here is an infinite number of problems of the form "Here's the answer, what's the question?" Usually such a setup is too wide open to be interesting. But a misfire problem, as I think I have defined it, has such a severely constrained answer that it is almost impossibly difficult. But "difficult" may be the wrong word. The trouble is that a misfire problem is a math problem with no implied search process whatever, except to review all the math and geometry that you already know. Instead of searching directly for a task last misfiring, say, for $n = 13$, it's much more likely that in the course of your mathematical ramblings you will someday bump into one.

I invite further contributions to the collection. Or arguments as to why this is too mushily stated a challenge.

Drawing de Bruijn Graphs

Herbert Taylor

The simplest kind of de Bruijn graph, in which the number of nodes is a power of two, has, at each node, two edges directed out and two edges directed in.

When the number of nodes is 2^n, we can label them with the integers from 0 to $2^n - 1$ and notice that they correspond to all the numbers expressible with n bits (n binary digits). Out from the node labeled x the two directed edges go to the nodes labeled $2x$ and $2x + 1$. If $2x$ or $2x + 1$ is bigger than $2^n - 1$, just subtract 2^n from it to bring the number back into the range from 0 to $2^n - 1$.

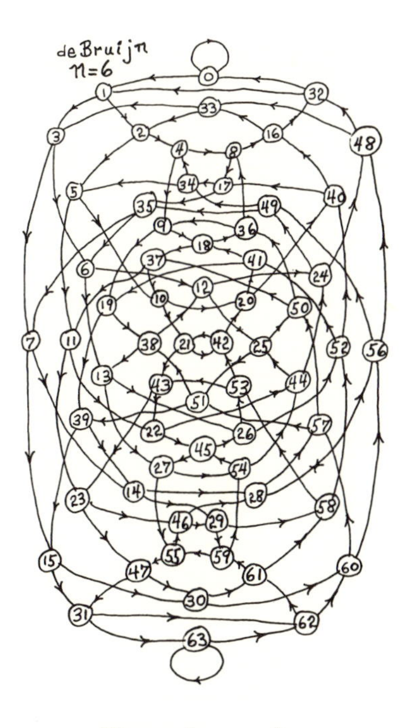

Figure 1. $n = 6$.

There is reason to think that the picture for $n = 6$ may fill a gap in the existing literature, because Hal Fredricksen told me he had not seen it in any book. Pictures up to $n = 5$ can be found in the classic "Shift Register Sequences" by Solomon W. Golomb, available from Aegean Park Press.

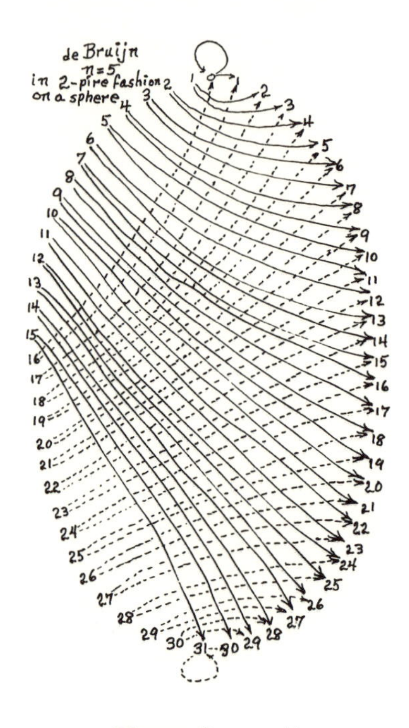

Figure 2. $n = 5$.

Bigger de Bruijn graphs get really hairy. The picture for $n = 5$ suggests 2-pire fashion[1] for drawing the de Bruijn graph on the sphere without any lines crossing each other. The scheme would be to put 0 on the North pole, $2^n - 1$ on the South pole, the numbers from 1 to $2^n - 2$ in order going south down the Greenwich Meridian, and down the international date line. Edges out from 0 to $2^{n-1} - 1$ on the meridian go east to the dateline, while edges out from 2^{n-1} to $2^n - 1$ go west.

[1]In 2-pire fashion each node of the graph is allowed to appear in two places in the picture. The earliest reference I know of to the *m*-pire (empire) problem is P. J. Heawood's 1891 paper in the *Quarterly Journal of Mathematics.*

See *Scientific American,* February 1980, Vol. 242, No. 2, "Mathematical Games," by Martin Gardner, pp. 14–22.

Computer Analysis of Sprouts

David Applegate, Guy Jacobson, and Daniel Sleator

Sprouts is a popular (at least in academic circles) two-person pen-and-paper game. It was invented in Cambridge in 1967 by Michael Patterson, then a graduate student, and John Horton Conway, a professor of mathematics. Most people (including us) learned about this game from Martin Gardner's "Mathematical Games" column in the July 1967 issue of *Scientific American.*

The initial position of the game consists of a number of points called *spots.* Players alternate connecting the spots by drawing curves between them, adding a new spot on each curve drawn. Each curve must be drawn on the paper without touching itself or any other curve or spot (except at end points). A single existing spot may serve as both endpoints of a curve. Furthermore, a spot may have a maximum of three parts of curves connecting to it. A player who cannot make a legal move loses. Shown below is a sample game of two-spot Sprouts, with the first player winning. Since draws are not possible, either the first player or the second player can always force a win, regardless of the opponent's strategy. Which of the players has this winning strategy depends on the number of initial spots.

Sprouts is an *impartial* game: The same set of moves is available to both players, and the last player to make a legal move wins. Impartial games have variants where the condition of victory is inverted: The winner is the player who cannot make a legal move. This is called the losing or *misère* version of the game.

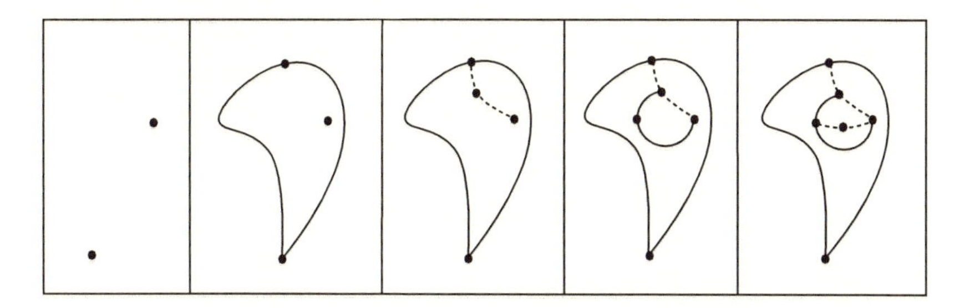

Figure 1. A sample game of two-spot Sprouts.

While Sprouts has very simple rules, positions can become fantastically complicated as the number of spots, n, grows. Each additional spot adds between two and three turns to the length of the game, and also increases the number of moves available at each turn significantly. Games with small numbers of spots can be (and have been) completely solved by hand, but as the number of spots increases, the complexity of the problem overwhelms human powers of analysis. The first proof that the first player loses in a six-spot game, performed by Denis Mollison (to win a 10-shilling bet!), ran to 47 pages.

Conway said that the analysis of seven-spot Sprouts would require a sophisticated computer program, and that the analysis of eight-spot Sprouts was far beyond the reach of present-day (1967!) computers. Of course, computers have come a long way since then. We have written a program that determines which player has a winning strategy in games of up to eleven spots, and in misère games of up to nine spots.

Our Sprouts Program

As far as we know, our program is the only successful automated Sprouts searcher in existence. After many hours of late-night hacking and experimenting with bad ideas, we managed to achieve sufficient time- and space-efficiency to solve the larger games. Our program is successful for several reasons:

- We developed a very terse representation for Sprouts positions. Our representation strives to keep only enough information for move generation. Many seemingly different Sprouts positions are really equivalent. The combination of this low-information representation and hashing (whereby the results of previous searches are cached) proved to be extremely powerful.
- Many sprouts positions that occur during the search are the *sum* of two or more non-interacting games. Sometimes it is possible to infer the value of the sum of two games given the values of the subgames. Our program makes use of these sum identities when evaluating normal Sprouts. These ideas are not nearly as useful in analyzing misère Sprouts. This is the principal reason that we are able to extend the analysis of the normal game further than the misère game.
- We used standard techniques to speed adversary search, such as cutting off the search as soon as the value is known, caching the results of previous searches in a hash table, and searching the successors of a position in order from lowest degree to highest.

- The size of the hash table turned out to be a major limitation of the program, and we devised and implemented two methods to save space without losing too much time efficiency. We discovered that saving only the losing positions reduced the space requirement by a large factor. To reduce the space still further we used a data compression technique.

Our Results

Here's what our program found:

Number of Spots	1	2	3	4	5	6	7	8	9	10	11
normal play	2	2	1	1	1	2	2*	2*	1*	1*	1*
misère play	1	2	2	2	1*	1*	2*	2*	2*		

A "1" means the first player to move has a winning strategy, a "2" means the second player has a winning strategy, and an asterisk indicates a new result obtained by our program.

The n-spot Sprouts positions evaluated so far fall into a remarkably simple pattern, characterized by the following conjecture:

Sprouts conjecture. *The first player has a winning strategy in n-spot Sprouts if and only if n is 3, 4, or 5 modulo 6.*

The data for misère Sprouts fit a similar pattern.

Misère sprouts conjecture. *The first player has a winning strategy in n-spot misère Sprouts if and only if n is 0 or 1 modulo 5.*

We are still left with the nagging problem of resolving a bet between two of the authors. Sleator believes in the Sprouts Conjecture and the Misère Sprouts Conjecture. Applegate doesn't believe in these conjectures, and he bet Sleator a six-pack of beer of the winner's choice that one of them would fail on some game up to 10 spots. The only remaining case required to resolve the bet is the 10-spot misère game. This problem seems to lie just beyond our program, our computational resources, and our ingenuity.[1]

---

[1] For a paper describing these results in more detail, go to http://www.cs.cmu.edu/˜sleator

Strange New Life Forms: Update

Bill Gosper

The Gathering for Gardner
Atlanta International Museum of Art and Design
Atlanta, Georgia

Dear Gathering,

Martin Gardner, by singlehandedly popularizing Conway's game of Life during the 1970s, sabotaged the Free World's computer industry beyond the wildest dreams of the KGB. Back when only the large corporations could afford computers because they cost hundreds of dollars an hour, clandestine Life programs spread like a virus, with human programmers as the vector. The toll in human productivity probably exceeded the loss in computer time.

With the great reduction in computation costs, and the solution of most of the questions initially posed by the game, it would be nice to report that, like smallpox, the Life bug no longer poses a threat. Sadly, this is not the case. While the Life programs distributed with personal computers are harmless toys whose infective power is 100% cancelled by TV, new developments in the game are still spreading through computer networks, infecting some of the world's best brains and machinery.

The biggest shock has been Dean Hickerson's transformation of the computer from simulation vehicle to automated seeker and explorer, leading to hundreds of unnatural, alien Life forms, of which the weirdest, as of late 1992, has to be

Time 0:

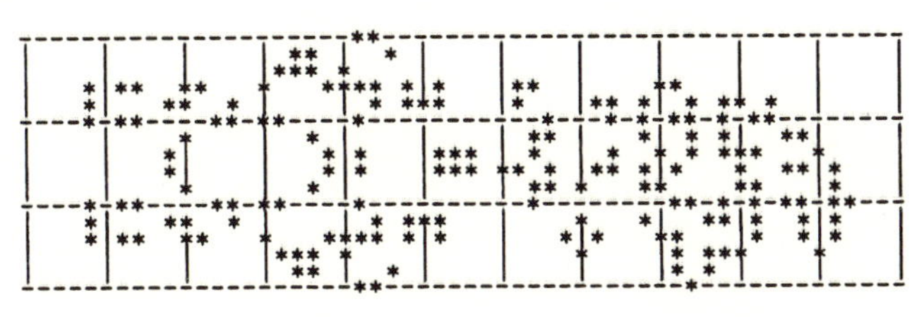

Note that it extends leftward with velocity $\frac{1}{4}$, with the left end (found by Hartmut Holzwart at the University of Stuttgart) repeating with period 4. But the (stationary) right end, found by Hickerson (at U.C. Davis) has period 5! Stretching between is a beam of darts that seems to move rightward at the speed of light. But if you interrupt it, the beam shortens in both directions at the speed of light, and then both ends explode.

Times 100–105:

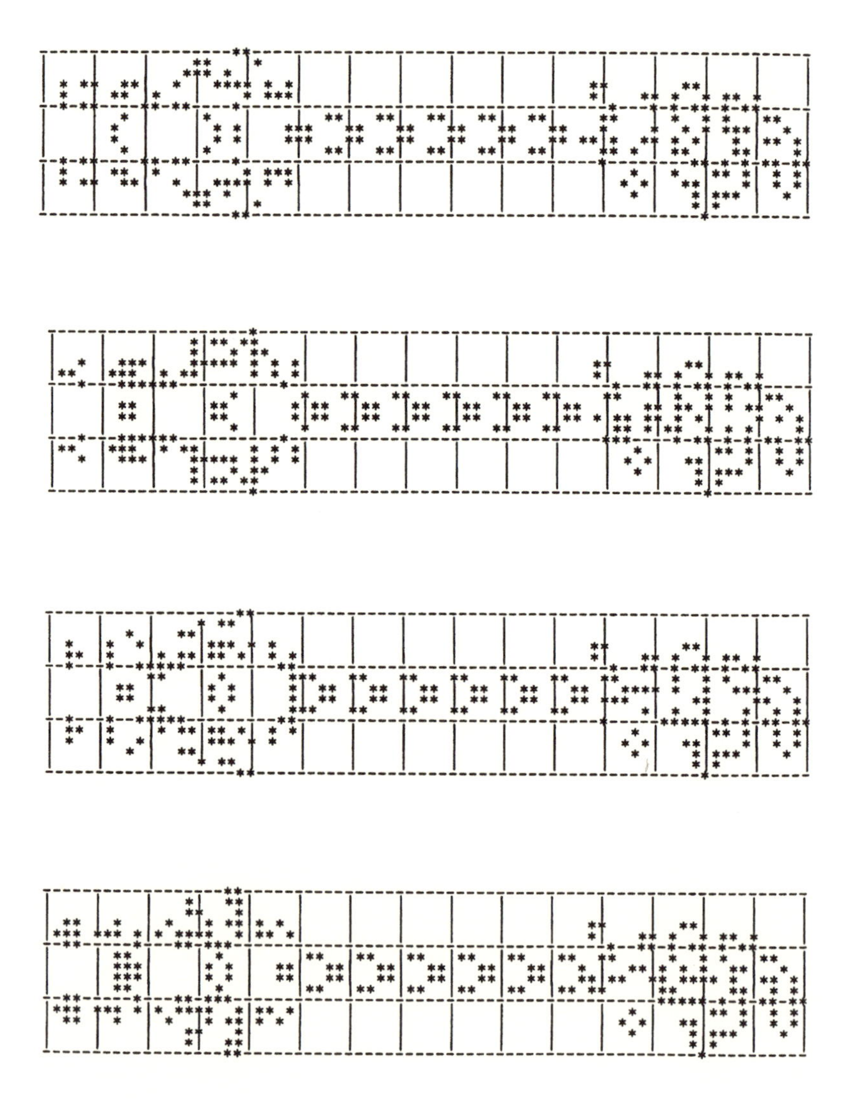

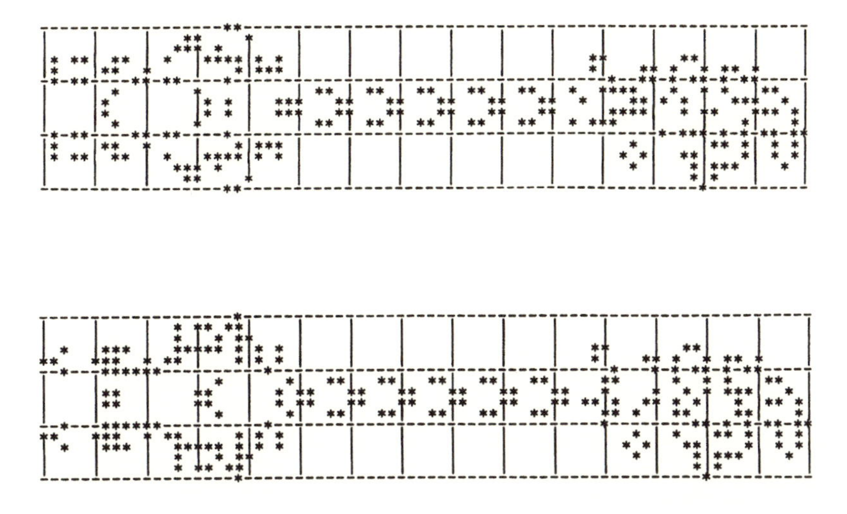

So, if you're wondering "Whatever happened to Life?," here is a synopsis of recent developments computer-mailed by Hickerson when the ineffable Creator of the Universe and its Laws himself asked the same question.

Hickerson's Synopsis

There have been developments in various areas by different people:

Oscillator and spaceship searches	by myself, David Bell, and Hartmut Holzwart
Construction of oscillators	mostly by David Buckingham and Bob Wainwright
Glider syntheses	mostly by Buckingham and Mark Niemiec
New glider guns	by Buckingham, Bill Gosper, and myself
Large constructions	mostly by Buckingham, Bill Gosper, Paul Callahan, and myself

Oscillator and Spaceship Searches. In 1989 I wrote a program to search for oscillators and spaceships. For a few months I ran it almost constantly, and I've run it occasionally since then. It found many oscillators of periods 3 and 4 (including some smaller than any previously known, and a period 4 resembling your initials, JHC!), a few oscillators with periods 5 (including some with useful sparks) and 6, infinitely many period 2, speed $c/2$ orthogonal spaceships, infinitely many period 3, speed $c/3$ orthogonal spaceships, one period 4, speed $c/4$ orthogonal spaceship, one period 4, speed $c/4$ diagonal spaceship (not the glider), and one period 5, speed $2c/5$ orthogonal spaceship. More recently, David Bell wrote a similar program that runs on faster machines. (Mine runs only on an Apple II.) Lately, he and Hartmut Holzwart have been producing oodles of orthogonal spaceships with speeds $c/2$, $c/3$, and $c/4$. Also, Hartmut found another with speed $2c/5$. Some of the $c/3$s and $c/4$s have sparks at the back which can do various things to $c/2$ spaceships that catch up with them. (These are used in some of the large patterns with unusual growth rates mentioned later.)

Construction of Oscillators. Due mostly to the work of Buckingham and Wainwright, we now have nontrivial examples (i.e., not just lcms of smaller period oscillators acting almost independently) of oscillators of periods $1-16, 18, 26, 28, 29, 30, 32, 36, 40, 44, 46, 47, 52, 54, 55, 56, 60, 72, 100$, 108, 128, $75+120n, 135+120n, 66+24n, 246+24n, 50+24n, 230+40n, 282+376n$, $846+376n, 136+8n, 150+5n$, and all multiples of $30, 44, 46, 94$, and 100. We also have an argument, based on construction universality, which implies that all sufficiently large periods are possible. (No doubt you figured that out for yourself long ago.)

The multiples of $30, 44, 46, 94$, and 100 and the $75+120n$ and $135+120n$ are based on gliders shuttling back and forth between either oscillators (or periods $15, 30, 46$, and 100) or output streams of glider guns (of periods $30, 44, 46, 94$, and $900+200n$). The $66+24n, 246+24n, 50+40n, 230+40n, 282+376n$, and $846+376n$ use a device discovered by Wainwright, which reflects a symmetric pair of parallel gliders; it consists of a still life ("eater3") and two spark producing oscillators, of periods 5, 6, or 47. (Any greater nonmultiple of 4 would also work, but 5, 6, and 47 are the only ones we know with the right sparks.) Periods $136+8n$ and $150+5n$ use a mechanism developed by Buckingham, in which a B heptomino is forced to turn a 90° corner. The turn can take either 64, 65, or 73 generations; by combining them we can build a closed loop whose length is any large multiple of 5 or 8; we put several copies of the B in the track to reduce the period to one of those mentioned. For example, I built a p155 oscillator in which 20 Bs travel around a track of length $3100(= 20 \times 64 + 28 \times 65)$ generations. For the multiples of 8, the 73 gen turn can emit a glider, so we also get glider guns of those periods.

Another amusing oscillator uses the glider crystallization that Bill mentioned in his centinal letter a few years ago. A period 150 gun fires toward a distant pair of pentadecathlons. The first glider to hit them is reflected 180° and collides with the second to form a honey farm. Subsequent gliders grow a crystal upstream; 11 gliders add a pair of beehives to the crystal. When the crystal reaches the guns, an eater stops its growth, and it begins to decay; two gliders delete one pair of beehives. When they're all gone, the process begins again.

Glider Syntheses. Rich has already mentioned this. I'll just add that some of Buckingham's syntheses of still lifes and billiard table oscillators are awe-inspiring.

New Glider Guns. In addition to the long-familiar period 30 and period 46 guns, there's now a fairly small period 44 gun that Buckingham built recently. The other guns are all pretty big. The largest by far that's actually been built is Bill's period 1100 gun (extensible to $900+200n$), based on the period 100 centinal; its bounding box is 13,584 by 12,112. Buckingham mercifully outmoded it with a comparatively tiny, centinal-based $500+200n$.

The period $136+8n$ guns based on Buckingham's B heptomino turns are also smaller. For example, there's a period $168 = (8 \times 73 + 4 \times 64)/5$ gun that's 119 by 119 and a period $136 = (40 \times 73 + 16 \times 64)/29$ gun that's 309 by 277.

Buckingham has also found extremely weird and elegant guns of periods 144 and 216, of size 149 by 149, based on a period 72 device discovered by Bob Wainwright.

The period 94 gun is 143 by 607; it's based on the "AK47" reaction discovered independently by Dave Buckingham and Rich Schroeppel. A honey farm starts to form but is modified by an eater and a block. It emits a glider, forms a traffic light, and then starts forming another honey farm in a different location. If you delete the traffic light, the cycle repeats every $47 \times 2 = 94$ generations. A close pair of AK47s can delete each other's traffic lights, so we can build a long row of them that is unstable at both ends. In the period 94 gun, two such rows, with a total of 36 AK47s, emit gliders that can crash to form MWSSs, which hit eaters, forming gliders that stabilize the ends of a row. Here's how to turn 2 MWSSs into a glider:

```
..O...............
O...O.............
.....O............
O....O............
.OOOOO............
..................
........O.........
......O...O.......
...........O......
......O....O......
.......OOOOO......
..................
..............OO..
..............O...
...............OOO
.................O
```

In *Winning Ways*, you (JHC) described how to thin a glider stream by kicking a glider back and forth between two streams. Using normal kickbacks, the resulting period must be a multiple of 8. I've found some eater-assisted kickbacks that give any even period. (Sadly, they don't work with period 30 streams. Fortunately, there are other ways to get any multiple of 30.)

It's also possible to build a gun that produces a glider stream of any period ≥ 15. The gun itself has a larger period; it uses various mechanisms to interleave larger period streams. It's fairly easy to get any period down to 18 this way (period 23 is especially simple), and Buckingham has found clever ways to get 15, 16, and 17. (I built a pseudo-period 15 gun based on his reactions; its bounding box is 373 by 372.)

New Year's Eve Newsflash: Buckingham has just announced ". . . I have a working construction to build a P14 glider stream by inserting a glider between two gliders 28 gens apart! . . . This will make it possible to produce glider streams of any period." (Less than 14 is physically impossible.)

Returning to Hickerson's account:

Many of the large constructions mentioned later require a period 120 gun. The smallest known uses a period 8 oscillator ("blocker") found by Wainwright to delete half the gliders from a period 60 stream. Here it's shown deleting a glider; if the glider is delayed by 60 (or, more generally, $8n + 4$) generations, it escapes:

```
.........O.
..........O
........OOO
...........
...........
...........
....OO.....
OO....OOOO.
OO..OO.OOO.
....O......
```

There are also combinations of two period 30 guns that give fairly small guns with periods 90, 120, 150, 10, 210, 300, and 360.

We can also synthesize light-, middle-, and heavyweight spaceships from gliders, so we have guns that produce these as well. For example, this reaction leads to a period 30 LWSS gun:

```
......O..
....O.O..
.OO..OO..
O.O......
..O.....O
......OO.
.......OO
```

Large Constructions. Universality implies the existence of Life patterns with various unusual properties. (At least they seem unusual, based on the small things we normally look at. Even infinite growth is rare for small patterns, although almost all large patterns grow to infinity.) But some of these properties can also be achieved without universality, so some of us have spent many hours putting together guns and puffers to produce various results. (Figuring out how to achieve a particular behavior makes a pleasant puzzle; actually building the thing is mostly tedious.)

The first such pattern (other than glider guns) was Bill's breeder, whose population grows like t^2. Smaller breeders are now known.

Probably the second such pattern is the "exponential aperiodic," versions of which were built independently by Bill and myself, and probably others. If you look at a finite region in a typical small pattern, it eventually becomes periodic. This is even true for guns, puffers, and breeders. But it's fairly easy to build a pattern for which this isn't true: Take two glider puffers, headed east and west, firing gliders southwest and northeast, respectively. Add a glider that travels northwest and southeast, using a kickback reaction each time it hits one of the glider waves. Cells along its flight path are occupied with decreasing frequency: The gap between the occupations increases by a factor of 3 or 9 each time, depending on which cells you're looking at.

Bill also built an "arithmetic aperiodic" in which the gap between occupations increases in an arithmetic progression.

Both of these patterns have populations tending to infinity. I've also built some aperiodic patterns with bounded populations. These use a glider salvo to push a block (or blinker); the reaction also sends back a glider (or two) which triggers the release of the next salvo.

In *Winning Ways* you (JHC) describe how to pull a block three units using two gliders and push it three units using 30. I found a way to pull it one unit with two gliders and push it one unit with three gliders. Using this I built a sliding block memory register, similar to the one you described. (It has one small difference: the "test for zero" is not a separate operation. Instead, a signal is produced whenever a decrement operation reduces the value to 0.)

Most of my large constructions are designed to achieve unusual population growth rates, such as $t^{1/2}, t^{3/2}, \log(t), \log(t)^2, t\log(t)$, and $t\log(t)^2$, and linear growth with an irrational growth rate. In addition, there are several "sawtooth" patterns, whose populations are unbounded but do not tend to infinity.

I also built a pattern (initial population $\leq$ 3000) that computes prime numbers: An LWSS is emitted in generation $120n$ if and only if n is prime. (This could be used to get population growth $t^2/\log(t)$, but I haven't built that.)

Paul Callahan and I independently proved that arbitrarily large puffer periods are possible. His construction is more efficient than mine, but harder to describe, so I'll just describe mine. Give a glider puffer of period N, we produce one of period $2N$ as follows: Arrange 3 period N puffers so their gliders crash to form a MWSS moving in the same direction as the puffers. The next time the gliders try to crash, there's already a MWSS in the way, so they can't produce another one. Instead, two of them destroy it and the third escapes; this happens every $2N$ generations. (Basically this mimics the way that a ternary reaction can double the period of a glider gun; the MWSS takes the place of the stable intermediary of the reaction.)

Dave Buckingham and Mark Niemiec built a binary serial adder, which adds two period 60 input streams and produces a period 60 output stream. (Of course, building such a thing from standard glider logic is straightforward, but they used some very clever ideas to do it more efficiently.)

–Dean Hickerson

To clarify, Buckingham's "awe-inspiring glider syntheses" are the constructions of prescribed, often large and delicate, sometimes even oscillating objects, entirely by crashing gliders together. The difficulty is comparable to stacking water balls in 1G. Warm ones. Following Dean's update, we were all astounded when Achim Flammenkamp of the University of Bielefeld revealed his prior discovery of Dean's smallest period 3 and 4 oscillators during an automated, months-long series of literally millions of random soup experiments. Thankfully, he recorded the conditions that led to these (and many other) discoveries, providing us with natural syntheses (and probably estimates) of rare objects.

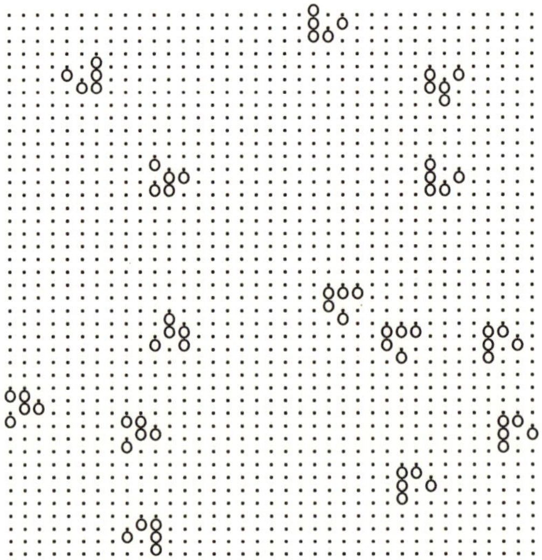

Conway called such oscillators billiard tables, and along with the rest of us, never imagined they could be made by colliding gliders.

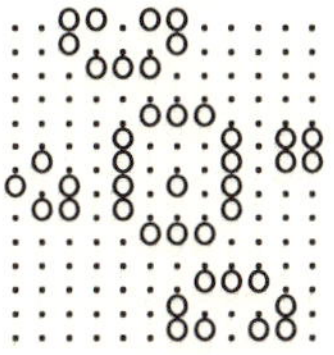

```
::8o.o8:::::
:::ȯȯȯ::::::
:::::ȯȯȯ::::
:ȯ::8:::8:88
ȯ88:8:ȯ:8:::
.ȯ8:8.ȯȯȯ:::
:::::::ȯȯȯ::
::::::8ȯ:ȯ8:
```

Finally, after a trial run of the foregoing around the Life net, Professor Harold McIntosh (of the Instituto de Ciencias at Puebla) responded: "The humor in that proposed introduction conjures up images of untold taxpayer dollars (or at least hours of computer time) disappearing into a bottomless black hole. That raises the question, 'Have other hours of computer time been spent more profitably?' (Nowadays, you can waste computer time all night long and nobody says anything.)

"Martin Gardner is a skilled presenter of ideas, and Life was an excellent idea for him to have the opportunity to present. Not to mention that the time was ideal; if all those computer hours hadn't been around to waste, on just that level of computer, perhaps the ideas wouldn't have prospered so well.

"In spite of Professor Conway's conjecture that almost any sufficiently complicated automaton is universal (maybe we can get back to that after vacations) if only its devotees pay it sufficient attention, nobody has yet come up with another automaton with anything like the logical intricacy of Life. So there is really something there which is worth studying.

"It might be worth mentioning the intellectual quality of Gardner's presentation of Life; of all the games, puzzles and tricks that made their appearance in his columns over the years, did anything excite nearly as much curiosity? (Well, there were flexagons.)

"Nor should it be overlooked that there is a more serious mathematical theory of automata, which certainly owes something to the work which has been performed on Life. Nor that there are things about Life, and other automata, that can be foreseen by the use of the theories that were stimulated by all the playing around that was done (some of it, at least)."

The Life you save may not fit on your disk.

–Bill Gosper

Many remarkable discoveries followed this writing. See, for example, `http://www.mindspring.com/~alanh/life/index.html`

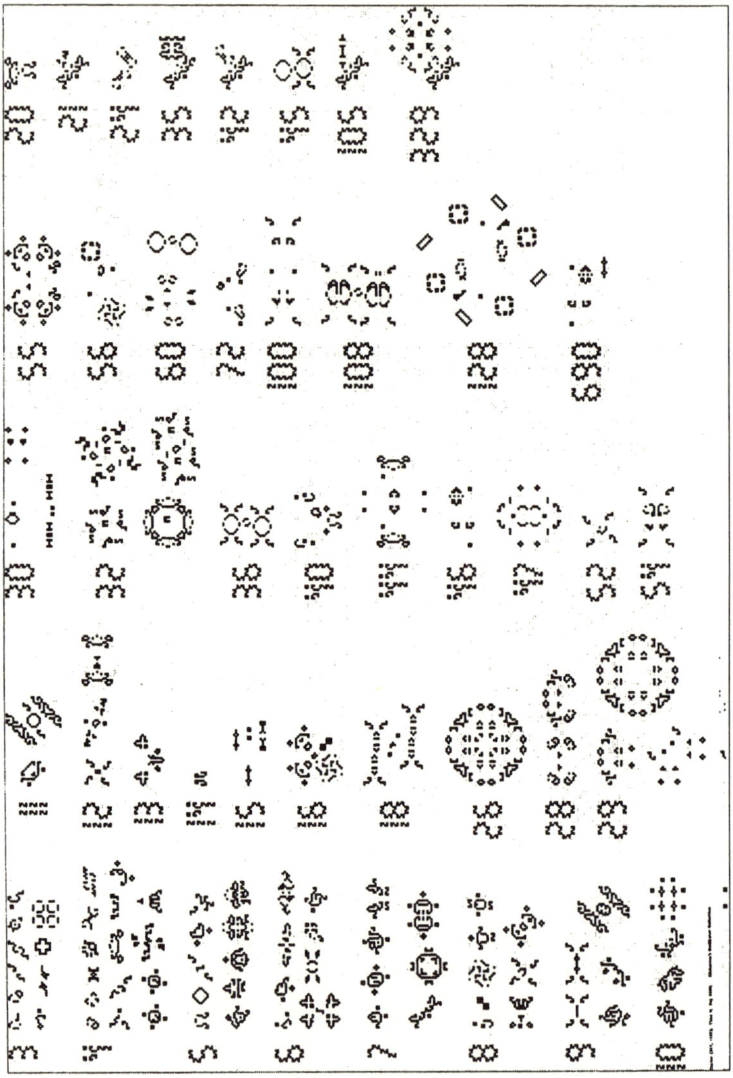

An early "stamp collection" by Dean Hickerson. These are representative oscillators of the (small) periods (indicated by the still-Life numerals) known by the end of 1992. As of the end of 1998, *all* periods have been found except 19, 23, 27, 31, 37, 38, 41, 43, 49, and 53.

Hollow Mazes

M. Oskar van Deventer

In October 1983 I discovered hollow mazes. A. K. Dewdney presented the concept in *Scientific American,* September 1988. In this article I shall give a more detailed description of the subject. The first part is about *multiple silhouettes* in general, the second part is about their application to *hollow mazes.*

Multiple Silhouettes

Silhouettes are often used in, e.g., mechanical engineering (working drawings) and robotics (pattern recognition). One particular mathematical problem reads: "Which solid object has a circular, a triangular, and a square silhouette?" The solution is the well-known *Wedge of Wallis* (Figure 1).

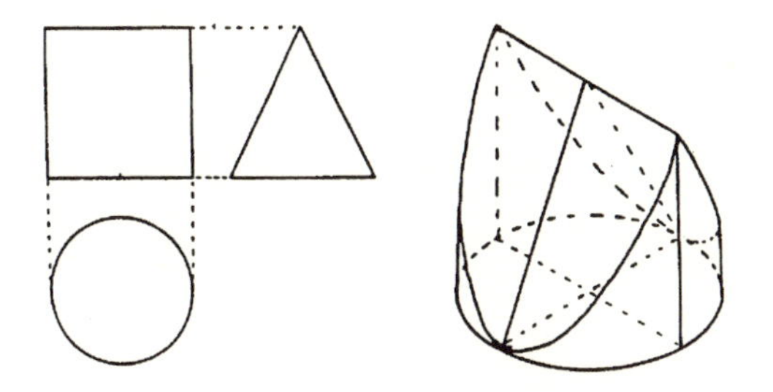

Figure 1. Wedge of Wallis.

A silhouette defines a cylindrical region in space, perpendicular to the silhouette surface. With this meaning the word no longer refers to an object, but only to a region of a flat surface (Figure 2).

Two or more silhouettes (not necessarily perpendicular to each other) define unambiguously a region or object in space that is the cross-section of two or more cylinders.

Dewdney introduced the term *projective cast* for an object defined by a number of silhouettes. Not every object can be defined as the projective

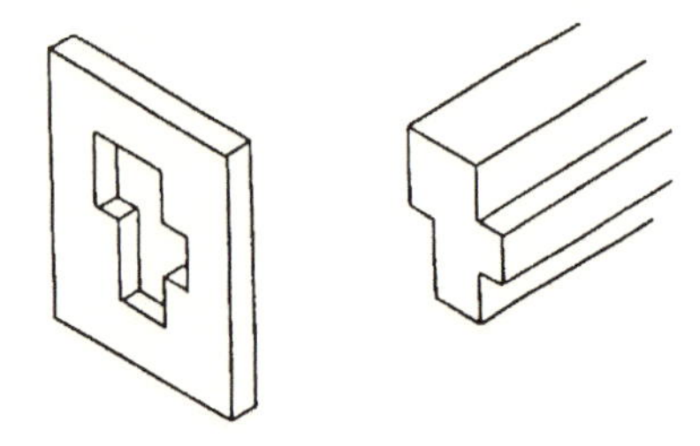

Figure 2. (a) Negative of a silhouette; (b) The cylindrical region in space defined by it.

cast of a (limited) number of silhouettes: One cannot cast a hollow cube or a solid sphere.

There are two interesting topological properties that apply to silhouettes, to projective casts, and to mazes and graphs in general. First, a graph may have either cycles or no cycles. Second, a graph may consist of either one part or multiple parts (Figure 3).

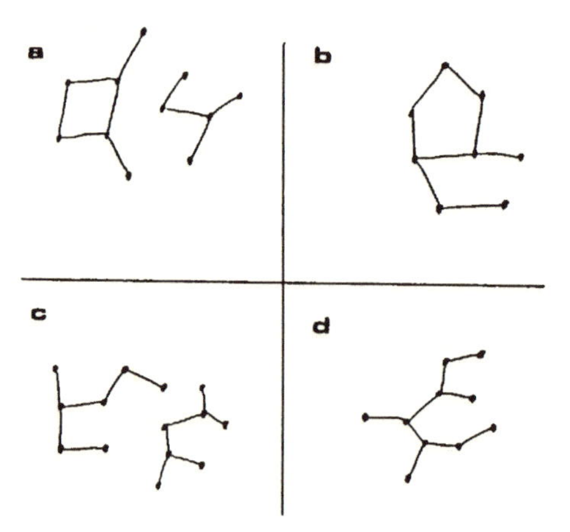

Figure 3. Topological properties of mazes and graphs: (a) multiple parts with cycles; (b) one part with cycles; (c) multiple parts, no cycles; (d) one part, no cycles.

Dewdney used the term *viable* for a silhouette or a maze that consists of one part and has no cycles. Even when silhouettes are viable, their projective cast can still have cycles (Figure 4) or multiple parts (Figure 5).

There is no simple connection between the properties of the silhouettes and the properties of their projective cast. The basic problem is the analysis

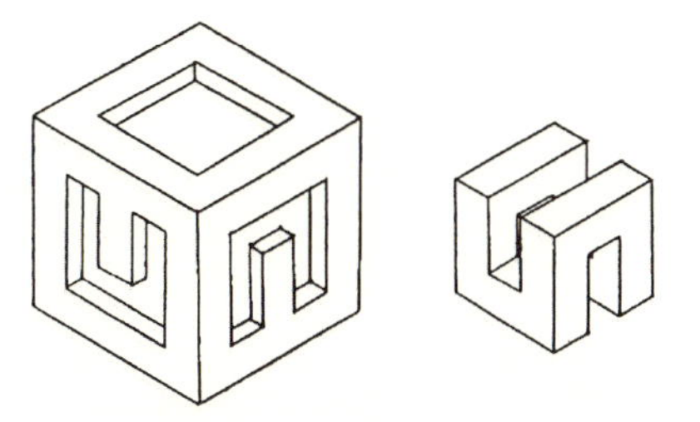

Figure 4. Three viable silhouettes yielding a projective cast with one cycle.

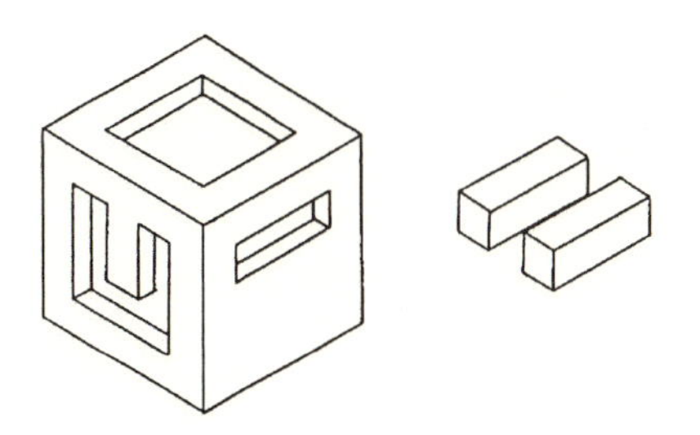

Figure 5. Three viable silhouettes yielding a projective cast with two parts.

and synthesis of silhouettes and projective casts: How can the properties of a projective cast be found without constructing it from its silhouettes, and how can a projective cast be given certain properties by its silhouettes? I do not know whether there is a general answer.

Hollow Mazes

A hollow maze is a rectangular box with six sides (Figure 6). Each side is a two-dimensional *control maze:* a surface into which slots have been cut. A cursor consisting of three mutually perpendicular spars registers one's position in the hollow maze. Each spar passes from one side of the box to the other, sliding along the slots of the control maze on each side. The two control mazes on opposite sides of the box are identical. In this way, a single three-dimensional maze is produced from three pairs of two-dimensional mazes. The resemblance between multiple silhouettes and hollow mazes is evident.

Control mazes are always viable because of their construction. There can be no cycles since the center of a cycle would fall out. Control mazes

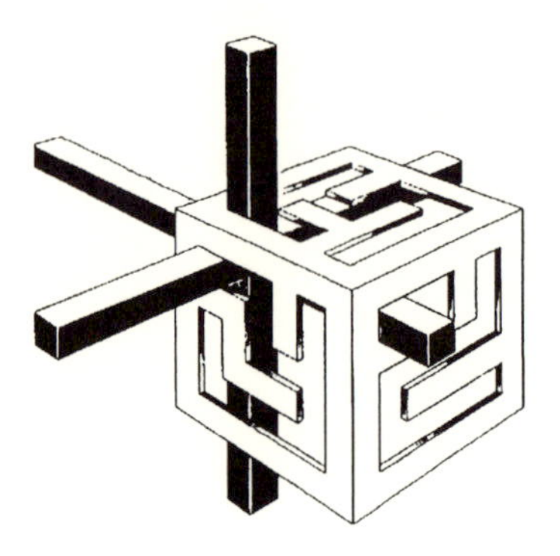

Figure 6. A simple hollow maze.

cannot consist of multiple parts, since a spar cannot jump from one part to another.

I have developed several restrictions for the investigation of hollow mazes:

- There are three (pairs of) perpendicular control mazes;
- the control mazes are viable;
- the control mazes are line mazes (i.e., their paths have zero width);
- a square or cubic grid is used for control mazes and their projective casts.

Now we can take another look at the topological properties of the projective cast. It seems that the projective cast cannot have cycles anymore, because the control mazes are line mazes. (I'm sure that there must be a simple proof of this, but I lack the mathematical tools.)

It can easily be tested whether or not a projective cast is viable, assuming that it cannot have cycles: We count the number of grid points of the projective cast. For example, a $3 \times 3 \times 3$ projective cast has 27 grid points. Then there must be exactly 26 "links" between these points for a viable cast. The number of links can be counted from the control mazes without constructing the projective cast.

Count the links in the hollow mazes of Figure 7. The dotted cross-section of Figure 7(a) has six links through it; the total number of links is found after taking all cross-sections.

There are several ways to construct a hollow maze, viable or not:

- **Trial and Error.** (Working at random.)

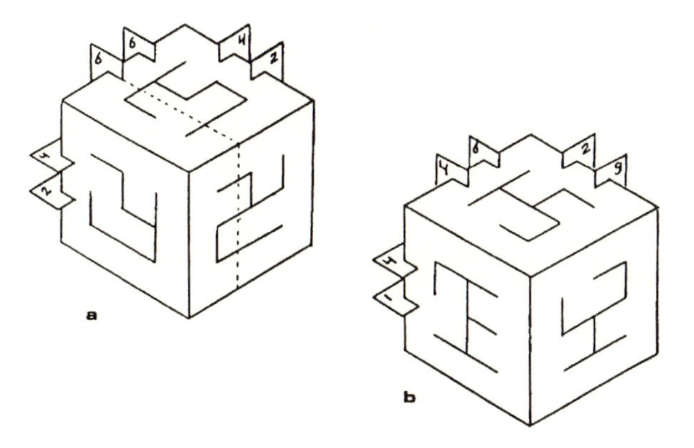

Figure 7. Test of the viability of a projective cast. (a) $2+4+6+6+4+2 < 27-1$; not viable. (b) $1+4+4+6+2+9 = 27-1$; viable!

- **Cut and Connect.** Cut the projective cast several times and link the pieces again (Figure 8). This method guarantees that the eventual hollow maze is viable. Helmut Honig, a German computer science student, showed me that not all viable hollow mazes can be constructed by this method. Figure 9 is a counterexample.

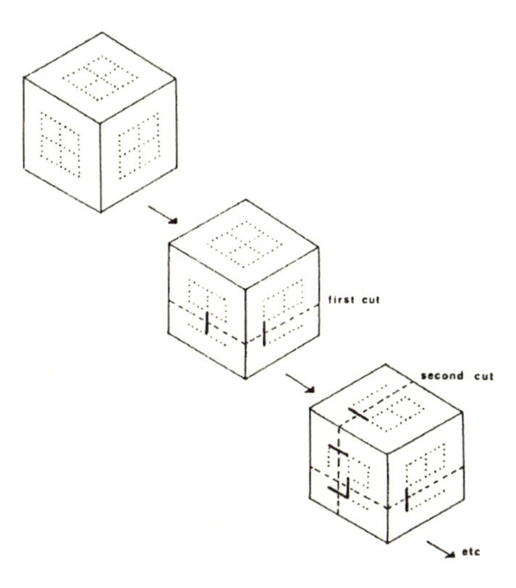

Figure 8. The cut and connect method for constructing hollow mazes.

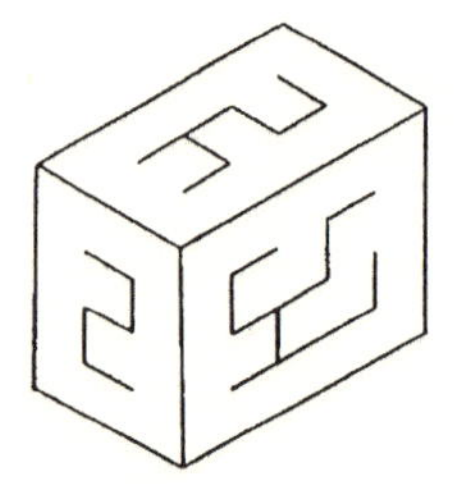

Figure 9. A viable hollow maze that cannot be constructed by the "cut and connect" method.

- **Regular Control Mazes.** See Figure 10. How should identical control mazes of this kind be placed to yield a viable hollow maze?
- **Recursion.** Replace each point by a hollow maze (see Figure 11).

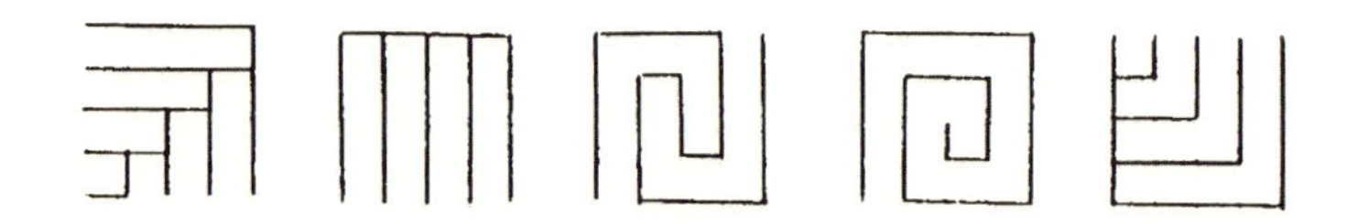

Figure 10. Regular control mazes.

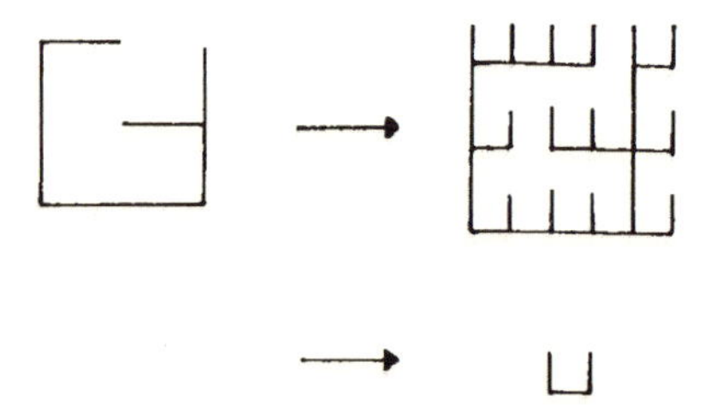

Figure 11. Construction of a hollow maze by recursion.

There is still much research to be done on multiple silhouettes and hollow mazes. I keep receiving letters from readers of *Scientific American*, and perhaps some of the problems mentioned above will be solved in the future.

Some Diophantine Recreations

David Singmaster

There are a number of recreational problems that lead to equations to be solved. Sometimes the difficulty is in setting up the equations, with the solution then being easy. Other times the equations are straightforward, but the solution requires some ingenuity, such as exploiting the symmetry of the problem. In a third class of problems, the equations and their solution are relatively straightforward, but there is an additional diophantine requirement that the data and/or the answers should be integral. In this case, it is not always straightforward to determine when the problem has solutions or to find them all or to determine the number of solutions. I discuss three examples of such diophantine recreations. The first is a simple age problem, done to illustrate the ideas. The second is the Ass and Mule Problem, for which I have just found a reasonably simple condition for integer data to produce an integer solution. The third is the problem of Selling Different Amounts at the Same Price Yielding the Same, for which I give an algorithm for finding all solutions and a new simple formula for the number of solutions.

A Simple Age Problem

Age problems have been popular since at least *The Greek Anthology* [16], compiled around A.D. 500 (although the date of this is uncertain, and the material goes back several centuries). The simplest are just problems of the 'aha' or 'heap' type, leading to one equation in one unknown. In *The Greek Anthology,* the problem of Diophantus' age gives us

$$\frac{x}{6}+\frac{x}{12}+\frac{x}{7}+5+\frac{x}{2}+4=x\,,$$

i.e., $75x/84+9=x$ or $9x/84=9$ or $x=84$. See Tropfke [20] for this and related problems.

In the late nineteenth century, complicated versions like "How Old Is Ann?" appeared, where the problem was in understanding the phrasing of the question, which contained statements such as "Mary is twice as old as Ann was when Mary was as old as Ann is now." (See Loyd [14])

Between these are a number of types of problem. The type that I want to examine is illustrated by the following, which is the earliest I have found, reportedly from *The American Tutor's Assistant,* 1791.[1]

When first the marriage knot was ty'd
Between my wife and me,
My age was to that of my bride
As three times three to three
But now when ten and half ten years,
We man and wife have been,
Her age to mine exactly bears,
As eight is to sixteen;
Now tell, I pray, from what I've said,
What were our ages when we wed?

A more typical, more comprehensible, albeit less poetic, version, but with the same numbers, appears at about the same time in Bonnycastle, 1824 [5].

> A person, at the time he was married, was 3 times as old as his wife; but after they had lived together 15 years, he was only twice as old; what were their ages on their wedding day?

We can state the general form of this problem as follows.

Problem (a, b, c); X is now a times as old as Y; after b years, X is c times as old as Y.

Thus the problem in *The American Tutor's Assistant* and Bonnycastle is Problem (3, 15, 2). It is easily seen that the problem leads to the equations

$$(1) \qquad X = aY; \qquad X + b = c(Y + b),$$

where X and Y denote the ages of X and Y. The solution of these is easily found to be

$$(2) \qquad Y = \frac{b(c-1)}{a-c}; \qquad X = aY = \frac{ab(c-1)}{a-c},$$

assuming $a \neq c$.

Now we give a diophantine aspect to our problem by asking the question: *If a, b, and c are integers, when are X and Y integers?* For example,

---

[1] I haven't actually seen the book; the problem is quoted in Bunt, et al. [6].

Problem (4, 3, 2) does not have an integral solution. Since X is integral if Y is, we can give a pretty simple answer: Y is integral if and only if

$$(3) \qquad a - c \quad \text{divides} \quad b(c-1)\,.$$

If we let g be the greatest common divisor (GCD) of $a - c$ and $c - 1$, denoted as $g = (a - c, c - 1)$, then (3) holds if and only if

$$(4) \qquad \frac{a-c}{g} \quad \text{divides} \quad b\,.$$

This gives us the kind of condition that allows us to construct all solutions. We pick arbitrary integers a and c, with $a > c > 1$, then let b be any multiple of $(a - c)/g$.

This pretty much settles the problem, although one can consider permitting some or all of the values to be rational numbers. Nonetheless, one can still get unexpected results, as in the following problem:

My daughter Jessica is 16 and very conscious of her age. Our neighbour Helen is just 8, and I teased Jessica by saying "Seven years ago, you were 9 times as old as Helen; six years ago, you were 5 times her age; four years ago, you were 3 times her age; and now you are only twice her age. If you are not careful, soon you'll be the same age!"

Jessica seemed a bit worried, and went off muttering. I saw her doing a lot of scribbling.

The next day, she said to me, "Dad, that's just the limit! By the way, did you ever consider when I would be half as old as Helen?" Now it was my turn to be worried, and I began muttering – "That can't be, you're always older than Helen."

"Don't be so positive," said Jessica, as she stomped off to school.

Can you help me out?

* * *

It is also possible to find a situation in which one person's age is an integral multiple of another's during six consecutive years.

Ass and Mule Problems

The classical ass and mule problem has the two animals carrying sacks. The mule says to the ass: "If you give me one of your sacks, I will have as many as you." The ass responds: "If you give me one of your sacks, I will have twice as many as you." How many sacks did they each have?

The general version of the problem for two individuals can be denoted $(a, b; c, d)$ for the situation where the first says: "If I had a from you, I'd have b times you," and the second responds: "And if I had c from you, I'd have d times you." Many versions of this problem occur, but it is traditional for the parameters a, b, c, d and the solutions, say x, y, to be integers. The general solution of the problem gives somewhat complicated expressions for x and y. Hence a natural question is to consider the diophantine question *Which integer problems have integer solutions?* In the early 1990s [1], I said I knew of no way to decide this that was simpler than actually seeing if the solutions were integers. Here I present a reasonably simple criterion that allows one to generate all the integer problems with integer solutions. I am indebted to S. Parameswaran for the letter that inspired this investigation.

The general problem $(a, b; c, d)$ leads to the system of equations

$$x + a = b(y - a)\,; \qquad y + c = d(x - c)\,. \tag{1}$$

The solutions are readily computed to be

$$x = c + \frac{(b+1)(a+c)}{bd-1}\,; \qquad y = a + \frac{(d+1)(a+c)}{bd-1}\,. \tag{2}$$

Thus, x and y are integers if and only if the second terms in (2) are integers. One can see from (2), and it is obvious from (1), that x is an integer if and only if y is an integer, so we only need to check one of the second terms in (2). This still doesn't give us a very simple or satisfactory criterion, however. Looking for a more symmetric solution, I first noted that $x + y$ has a symmetric expression, and I thought that the integrality of this was equivalent to integrality of x and y; unfortunately, this doesn't hold. Luckily, though, this false trail led me to the following simple result, which is similar to, but somewhat more complex than, the age problem.

The values of x and y are integers if and only if $bd - 1$ divides both $(b + 1)(a + c)$ and $(d + 1)(a + c)$, which is if and only if $bd - 1$ divides $\mathrm{GCD}[(b+1)(a+c), (d+1)(a+c)] = (a+c)(b+1, d+1)$, where $(b+1, d+1)$ denotes the GCD of $b + 1$ and $d + 1$, as before.

Consider $g = (b+1, d+1)$. This g also divides $(b+1)(d+1) - (b+1) - (d+1) = bd - 1$. Now the last statement of the previous paragraph can be divided by g to give us that x and y are integers if and only if

$$\frac{bd-1}{(b+1, d+1)} \quad \text{divides} \quad a + c\,. \tag{3}$$

This seems to be as simple a criterion for integrality as one could expect. The criterion allows us to pick arbitrary b and d, assuming $bd \neq 1$, and then determines which values of a and c give integral solutions. I find it particularly striking that a and c only enter via the sum $a + c$. I am also intrigued to see that any b and any d can be used, assuming $bd \neq 1$, which I would not have predicted.

The basic role of $b+1$ and $d+1$ in the above leads one to ask whether the problem can be recast in some way to make these the natural parameters, but doing so doesn't seem to make the result significantly clearer.

Although the problem traditionally has all integral parameters, this is not really essential; the admission of irrational values, however, loses all number-theoretic interest. One can deal with rational a, c, x, y by scaling the problem so that one has an integer problem, so it seems most interesting to assume that we have integral a and c and we want integral x and y. It is then quite feasible to consider rational b and d. Setting $b = \beta/\alpha$, $d = \delta/\alpha$, where GCD $(a, \beta, \delta) = 1$, it is direct to show that x and y are integral if and only if

$$(4) \qquad \frac{\beta\delta - \alpha^2}{(\beta + \alpha, d + a)} \quad \text{divides} \quad \alpha(a + c).$$

Dr. Parameswaran interpreted the problem as though the second statement was also made by the first animal. Thus he considered the problem where equations (1) are replaced by

$$(6) \qquad x + a = b(y - a)\,; \qquad\qquad x + c = d(y - c)\,.$$

This is easily solved to obtain

$$(7) \quad y = a + \frac{(d+1)(a-c)}{b-d}\,; \qquad\qquad x = \,-a + \frac{b(d+1)(a-c)}{b-d}\,.$$

The expression for x is a bit complicated, but we see from (6) that x is an integer if y is an integer, though the converse may fail. From the first equation, we get a fairly direct way to check whether y is integral, but by analogy with the argument above, we see that y is integral if and only if $b-d$ divides $(d+1)(a-c)$. Now consider $g = (b-d, d+1)$. We easily see that $g = (b+1, d+1)$, so Parameswaran's version has integral solutions if and only if

$$(8) \qquad \frac{b-d}{(b+1, d+1)} \quad \text{divides} \quad a - c\,.$$

From this, one can easily generate all examples.

Thinking about Parameswaran's version led me to wonder if there was any case where both versions gave integral answers, possibly even the same answers. That is, suppose you can hear the animals but can't tell which one is speaking, so you don't know whether the statements are made by the same or different animals. Are there problems where the answers are integral, or even the same, in either case? Perhaps surprisingly, there are problems where the answers are indeed the same, and I leave it for the reader to discover them.

History of the Ass and Mule Problem. The earliest known version of this problem has an ass and a mule with parameters (1, 2; 1, 1) and is attributed to Euclid, from about 300 B.C. Heiberg's edition [10] gives the problem in Greek and Latin verses.

Diophantus studied the problem in general. He proposes "to find two numbers such that each after receiving from the other may bear to the remainder a given ratio." He gives (30, 2; 50, 3) as an example. He also covers similar problems with three and four persons.

Two versions, (10, 3; 10, 5) and (2, 2; 2, 4), appear in *The Greek Anthology* [16]. Alcuin gives an unusual variant of the problem in that he assumes the second person starts from the situation after the transfer mentioned by the first person takes place. That is, the second equation of (1) would be replaced by

$$y - a + c = d(x + a - c)\,.$$

This is our problem $(a, b; c - a, d)$.

By about the ninth century, the problem was a standard, not only in Europe but also in India and the Arab world, and it has remained a standard problem ever since.

The problem is readily generalised to more than two participants, but then there are two forms, depending on whether 'you' refers to all the others or just the next in cyclic sequence. I denote these as I$-(a, b; c, d; \ldots)$ for the case where 'you' means all the others and II$-(a, b; c, d; \ldots)$ for the case where 'you' means just the next person in a cyclic sequence. The earliest three-person version that I know of appears in India, about A.D. 850, in the work of Mahavira [15]; the problem is of type I. The earliest four-person version I have seen is Arabic, in al-Karkhi (aka al-Karaghi) from about 1010 [13], and the problem is of type II.

Tropfke [20] discusses the problem and cites some Chinese, Indian, Arabic, Byzantine, and Western sources. The examples cited in the Chiu Chang Suan Shu [7] state, e.g., "If I had half of what you have, I'd have 50." This is a different type of problem and belongs in Tropfke's previous section. The other Chinese reference is a work of about A.D. 485 that has only been translated into Russian and may well be similar to the earlier Chinese example, so that, surprisingly, the Ass and Mule problem may not be in any Chinese source.

Below I give a summary of versions that I have noted up through Fibonacci, who typically gives many versions, including some with five persons, an inconsistent example, and extended versions where a person says, e.g.,

"If you give me 7, I'd have 5 times you plus 1 more." This includes all the relevant sources cited by Tropfke except Abu al-Wafa (only available in Arabic) and "Abraham," who is elsewhere listed as probably being from the early fourteenth century and hence after Fibonacci, although the material is believed to be based on older Arabic sources.

(1, 2; 1, 1)	c. 300 B.C.	Euclid
(30, 2; 50, 3)	c. 250	Diophantus
(10, 3; 10, 5)	c. 510	*The Greek Anthology*
(2, 2; 2, 4)	c. 510	*The Greek Anthology*
(2, 1; 0, 2)	9th century	Alcuin
I-(9, 2; 10, 3; 11,5)	c. 850	Mahavira
I-(25, 3; 23, 5; 22, 7)	c. 850	Mahavira
II-(1, 2; 2, 3; 3, 4; 4, 5)	c. 1010	al-Karkhi
(100, 2; 10, 6)	1150	Bhaskara
(1, 1; 1, 10)	1202?	Fibonacci
(7, 5; 5, 7)	1202?	Fibonacci
I-(7, 5; 9, 6; 11, 7)	1202?	Fibonacci, with the problem statement giving the 6 as a 7
I-(7, 3; 8, 4; 9, 5; 11, 6)	1202?	Fibonacci, noting that this system is inconsistent
I-(7, 2; 8, 3; 9, 4; 10, 5; 11, 6)	1202?	Fibonacci, with the problem statement giving the 6 as a 7
I-(7, 4; 9, 5; 11, 6)	1202?	Fibonacci
I-(7, 3; 9, 4; 11, 5)	1202?	Fibonacci, done two ways

Selling Different Amounts at the Same Price Yielding the Same

Although this problem is quite old, there seems to be no common name for it, hence the above title, which is rather a mouthful. The problem is treated in one form by Mahavira [15], Sridhara [18], and Bhaskara [4], but this form is less clearly expressed and has infinitely many solutions, so I will first describe the later form which is first found in Fibonacci [11]. I quote a version that appears as No. 50 in the first printed English riddle collection "The Demaundes Joyous," printed by Wynken de Worde in 1511 [9].

> A man had three daughters of three ages; which daughters he delivered, to sell, certain apples. And he took to the eldest daughter fifty apples, and to the second thirty apples, and to the youngest ten apples; and all these three sold in like many for a penny, and brought home in like such money. Now how many sold each of them for a penny?

'In like many' means 'the same number' and 'in like much money' means 'the same sum of money.' Anyone meeting this problem for the first time soon realises there must be some trick to it. Fibonacci has no truck with trick questions and gives the necessary information – the sellers sell part of their stock at one price and then the remainders at another price. This may happen by the sellers going to different markets, or simply changing prices after lunch, or lowering prices late in the day in an attempt to sell their remainders, or raising them late in the day when few items are left and new buyers arrive. Although the objects are usually eggs, or sometimes fruit, which normally are all of the same value, Tartaglia [19] justifies the different prices by having large and small pearls. Even with this trick exposed, it can take a fair amount of time to work out a solution, and it is not immediately obvious how to find or count all the solutions.

Let us denote the problem with numbers $c_1, c_2, \ldots$ by $(c_1, c_2, \ldots)$, so that the above is (10, 30, 50). This is historically by far the most common version. Fibonacci only gives two-person versions: (10, 30); (12, 32); (12, 33). Fibonacci is one of the few to give more than a single solution. He gives a fairly general rule for generating solutions [12], which would produce the positive solutions with the smaller price equal to 1, and he gives five solutions for his first version – but there are 55, of which 36 are positive. Fibonacci goes on to consider solutions with certain properties – e.g., if the prices are also given, is there a solution; find all solutions with one price fixed; find solutions where the amounts received are in some ratio, such as the first receives twice the second.

The question of finding all solutions for a three-person problem seems to have first been tackled in the 1600s. According to Glaisher [12], the 1612 and 1624 editions of Bachet [3][2] give a fairly general rule for this case and apply it to (20, 30, 40). In 1874, Labosne revised the material, dropping Bachet's rule and replacing it with some rather vague algebra; he added several more versions as well, for which he often gives fractional answers, distinctly against the spirit of the problem. The 1708 English edition of Ozanam [17], which is essentially the same as the first edition of 1694, considers (10, 25, 30) and gives two solutions. The extensively revised and enlarged edition of 1725 (or 1723) outlines a fairly general method and correctly says there are ten solutions, contrary to De Lagny's statement that there are six solutions. Neither Glaisher nor I have seen the relevant work of De Lagny, but we both conjecture that he was counting the positive solutions, of which there are indeed six.

During the 1980s, I tried several times to find an easy way to generate the solutions. I have now arrived at a reasonably simple approach that turns

[2] I have not seen these editions.

out to be essentially a generalisation and extension of Ozanam's method. Glaisher's paper contains most of the ideas involved, but he is so verbose and gives so many pages of tabular examples, special cases, and generalizations that it is often difficult to see where he is going – it is not until the 51st page that he gives the solutions for the problem (10, 30, 50), which he starts considering on the first page. He also assumes that the c_i are an arithmetic progression, but it is not until late in the paper that one sees that he solves versions that are not in arithmetic progression by the simple expedient of inserting extra values. By (4) below, this doesn't change the problem, but I prefer to deal with the original values. Ayyangar [2] gives a much simplified general method for the Indian version of the problem and says many of his solutions are not given by Glaisher's method on page 19 – although I can't tell if Glaisher intends this to be a complete solution. Ayyangar's method is quite similar to my method, so I later sketch his method in my notation and fill in a gap.

Most previous methods yielded solutions involving some choices, subject to some conditions, but did not thoroughly check that all solutions are obtained. No one else has noticed that the number of solutions in the Western version can be readily computed. Indeed, whenever Glaisher gives the number of solutions, he displays or indicates all of them.

Consider the problem $(c_1, c_2, \dots, c_n)$, i.e., the ith person has c_i items to sell. We can reorder them so that $c_1 < c_2 < \cdots < c_n$, and we assume $n > 1$ to avoid triviality. Let a_i be the number of items that the ith person sells at the higher price A, and let b_i be the number of items that the ith person sells at the lower price B. Then we have the following:

$$a_i + b_i = c_i\,; \tag{1}$$

$$Aa_i + Bb_i = C\,, \tag{2}$$

where C is the constant amount received. We have $2n + 3$ unknowns connected by these $2n$ equations. We thus expect a three-dimensional solution set. Normally the objects being sold are indivisible, so we want the numbers a_i, b_i, c_i to be integers.

Because of the symmetry between A and B, we can assume and have assumed $A > B$. This makes $a_1 > a_2 > \cdots > a_n$ and $b_1 < b_2 < \cdots < b_n$. (One can reverse A and B in order to make the a_is increase, but the amounts sold at the higher price are the smaller numbers and hence are easier to deal with.)

From $Aa_1 + Bb_1 = Aa_2 + Bb_2$, we get $A(a_1 - a_2) = B(b_2 - b_1)$. Hence A/B is a rational number, and we can assume that A and B are integers with no common factor, i.e., that $(A, B) = \text{GCD}(A, B) = 1$. This is equivalent to scaling the prices in equations (2) and reduces the dimensionality of the solution set to two. This is quite reasonable, as solutions that differ only in price scale are really the same. Solutions in the literature, however, are often given with fractional values, in particular $B = 1/\beta$ for some integer β – e.g., the most common solution of the (10, 30, 50) problem has $A = 3, B = 1/7$ – and the usual solutions have each person selling as many groups of β as possible. In the present approach, these prices are considered the same as $A = 21, B = 1$, and they are considered to give the same sets of solutions. Glaisher considers $A = 3$, $B = 1/7$ with the condition that C be integral, which restricts the solutions to a subset of those for the case $A = 21, B = 1$. In some versions of the problem, the yield C is specified. This is also equivalent to scaling the prices, and the solution set is again two-dimensional, unless one imposes some further condition such as B being the reciprocal of an integer.

Now use equation (1) to eliminate b_i from (2), giving us

$$(A - B)a_i = C - Bc_i\,. \tag{3}$$

Subtracting each of these from the case $i = 1$ leaves us $n - 1$ equations:

$$(A - B)(a_1 - a_i) = B(c_i - c_1)\,. \tag{4}$$

Since $(A, B) = 1$ implies $(A - B, B) = 1$, it follows that (4) has an integral solution for $a_1 - a_i$ if and only if $A - B$ divides $c_i - c_1$ and this holds for each i if and only if

$$A - B \quad \text{divides} \quad \mathrm{GCD}(c_i - c_1)\,. \tag{5}$$

It is now convenient to adopt the following:

$$\alpha_i = a_1 - a_i\,; \qquad \gamma_i = c_i - c_1\,; \qquad \Gamma = \mathrm{GCD}(\gamma_i)\,.$$

Thus (5) can be rewritten as $A - B$ divides Γ. This tells us that $d(A - B) = \Gamma$ for some d, and (4) gives us $a_i = Bd\gamma_i/\Gamma$, so that

$$\alpha_i \quad \text{is a multiple of} \quad \gamma_i/\Gamma\,. \tag{6}$$

Now we will take α_2 as the first basic parameter of our solution, and we note that $1 \leqslant \alpha_2 \leqslant c_1$. From (6), we must have γ_2/Γ dividing α_2, and I claim that any such α_2 generates an integer solution of our problem. From the case $i = 2$ of (4), we have $(A - B)\alpha_2 = B\gamma_2$. Let $\gamma = (\alpha_2, \gamma_2)$, so the previous equation becomes $(A - B)\alpha_2/\gamma = B\gamma_2/\gamma$, where $(A - B, B) = (\alpha_2/\gamma, \gamma_2/\gamma) = 1$. Hence $A - B = \gamma_2/\gamma$, $B = \alpha_2/\gamma$. We also get $A = (\alpha_2 + \gamma_2)/\gamma$, $A/B = (\alpha_2 + \gamma_2)/\alpha_2 = 1 + \gamma_2/\alpha_2$. From just before (6), we have $\Gamma\alpha_2 = Bd\gamma_2$, so $\Gamma\alpha_2/ = Bd\gamma_2/\gamma$. Since $(\alpha_2/\gamma, \gamma_2/\gamma) = 1$, this implies that γ_2/γ divides Γ, i.e., $A - B$ divides Γ, which is the condition (5) for all the equations (4) to have integral solutions.

As an example, consider the problem discussed in Ozanam, namely (10, 25, 30). Then $\gamma_2 = 15$, $\gamma_3 = 20$, $\Gamma = 5$, and we choose α_2 as a multiple of $\gamma_2/\Gamma = 3$.

For $\alpha_2 = 3$, we have $\gamma = (3, 15) = 3$, $A - B = 5$, $B = 1$, $A = 6$, and there are solutions with $\alpha_i = \gamma_i/5$, i.e., $a_1 - a_2 = 3$, $a_1 - a_3 = 4$, hence $a_2 - a_3 = 1$. We can now let a_1 vary as the second basic parameter of our solution and we can let $a_1 = 10, 9, 8, 7, 6, 5, 4$, giving us seven solutions.

For $\alpha_2 = 6$, we have $\gamma = (6, 15) = 3$, $A - B = 5$, $B = 2$, $A = 7$, and there are solutions with $\alpha_i = 2\gamma_i/5$, i.e., $a_1 - a_2 = 6$, $a_1 - a_3 = 8$, hence $a_2 - a_3 = 2$. We can now let $a_1 = 10, 9, 8$, giving us three solutions.

For $\alpha_2 = 9$, we get no solutions.

Thus Ozanam is correct, but four of these solutions have either $b_1 = 0$ or $a_3 = 0$, so there are just six positive solutions.

Counting the Solutions. We get solutions in the above section for each α_2 that is a multiple of γ_2/Γ and in the interval $1 \leqslant \alpha_2 \leqslant c_1$. For such an α_2, we can let $a_1 = c_1, c_1 - 1, \ldots$, until a_n becomes zero. From the case $i = n$ of (4), we have $(A - B)(a_1 - a_n) = B\gamma_n$, so that $a_n = 0$ when

$$a_1 = \frac{B\gamma_n}{A - B} = \frac{\alpha_2\gamma_n}{\gamma_2}. \tag{7}$$

(This expression is an integer, since $A - B$ divides γ_n.) Hence, for a given α_2, there are

$$c_1 - \frac{\alpha_2\gamma_n}{\gamma_2} + 1 \quad \text{solutions.} \tag{8}$$

In the above example, $\alpha_2 = 3$ gives $10 - 3 \times 20/15 + 1 = 7$ solutions; $\alpha_2 = 6$ gives $10 - 6 \times 20/15 + 1 = 3$, and $\alpha_2 = 9$ gives a negative value, indicating no solutions.

We can now determine the possible values of α_2. From the case $i = n$ of (4), we have

$$\alpha_n = \frac{B\gamma_n}{A - B} = \frac{\alpha_2\gamma_n}{\gamma_2}. \tag{9}$$

Now the largest value α_n can have is c_1, so we have

$$1 \leqslant \alpha_2 \leqslant \frac{\gamma_2 c_1}{\gamma_n}. \tag{10}$$

In our example, we find $1 \leqslant \alpha_2 \leqslant 7.5$.

Recalling that α_2 is a multiple of γ_2/Γ, set

$$\alpha_2 = \frac{k\gamma_2}{\Gamma}. \tag{11}$$

Then (10) becomes

$$1 \leqslant k \leqslant \frac{\Gamma c_1}{\gamma_n}. \tag{12}$$

The upper bound may not be an integer, so we let

$$K = \left[\frac{\Gamma c_1}{\gamma - n}\right], \qquad \text{where [ ] is the greatest integer function.} \tag{13}$$

Substituting (11) into (8) and summing gives us the number of solutions,

$$N = \sum_{k=1}^{K}(c_1 - \frac{k\gamma_n}{\Gamma} + 1) = K(c_1 + 1) - \frac{\gamma_n K(K+1)}{2\Gamma}. \tag{14}$$

Again, in our example, we have $K = 2$ and

$$N = 2 \times 11 - 20 \times 2 \times 3/2 \times 5 = 10.$$

Most of the earlier writers excluded zero values and only considered positive solutions. We can count the number N_+ of positive solutions by finding the number of solutions with a zero value. These occur when $b_1 = 0$, i.e., $a_1 = c_1$, and when $a_n = 0$, i.e., $a_1 = \alpha_2\gamma_n/\gamma_2$ by (7). Hence we normally get $2K$ situations with a zero value, unless it happens that $c_1 = \alpha_2\gamma_n/\gamma_2$, when two zero values occur in the same solution. Rearranging, we find that this occurs when $k = \Gamma c_1/\gamma_n$, i.e., when the upper bound for k given in (13) is an exact integer. In that case, we have $2K - 1$ solutions with zero values. Subtracting $2K$ or $2K - 1$ from N yields N_+.

Looking again at our example, $N_+ = 10 - 2 \times 2 = 6$, perhaps as intended by De Lagny.

Most of the examples in the literature have the c_i in arithmetic progression (AP). A little thought shows that then Γ is the common difference of the progression, so we can write $c_i = c_1 + (i-1)\Gamma$. Then we have $\gamma_n = (n-1)\Gamma$; $\gamma_n/\Gamma = n - 1$:

$$K = \left[\frac{c_1}{n-1}\right]; \qquad N = K(c_1 + 1) - \frac{(n-1)K(K+1)}{2}. \tag{15}$$

A number of early cases have $n = 2$, which is trivially an AP. We then have $K = c_1$; $N = c_1(c_1 + 1)/2$; $N_+ = (c_1 - 2)(c_1 - 1)/2$.

The majority of examples are APs with $n = 3$. All but one of these have c_1 even, and then $K = c_1/2$; $N = c_1^2/4$; $N_+ = (c_1 - 2)^2/4$. If c_1 is odd, then we get $K = (c_1 - 1)/2$; $N = (c_1^2 - 1)/4$; $N_+ = [(c_1 - 2)^2 - 1]/4$.

A few longer APs occur — I have seen examples with $n = 5, 7, 9$. A few cases occur that are not APs (see below).

Tartaglia's problem 139 is the one with two sizes of pearls and is (10, 20, ... , 90). From (15), we have $K = 1$; $N = 3$; $N_+ = 1$. This led me to ask when a problem admits only one solution or only one positive solution. Glaisher [12] also studied this problem and was intrigued by this observation, noting the analogous cases with 10 and 11 people.

In general, $N = 1$ implies that the only solution occurs with $k = 1$ and has $a_1 = c_1$ and $a_n = 0$, which gives us $k = \Gamma c_1/\gamma_n$, hence $\gamma_n = \Gamma c_1$ or $c_n = (\Gamma + 1)c_1$. For an AP, this is if and only if $c_1 = n - 1$. For example, (10, 20, ... , 110) has $N = 1$.

Similarly, $N_+ = 1$ implies that the unique positive solution occurs with $k = 1$ and has $a_1 = c_1 - 1$ and $a_n = 1$. A little manipulation shows that this works if and only if

$$c_n = (\Gamma + 1)c_1 - 2\Gamma \,. \tag{16}$$

For an AP, this turns out to be if and only if $c_1 = n + 1$, as in Tartaglia's example.

Initially I thought that $N_+ = 1$ would imply that $N = 3$, but when $n = 2$, we find $K = 3$ and $N = 6$, while when $n = 3$, we have $K = 2$ and $N = 4$. We will get $N = 3$ only if $K = 1$, i.e., $\Gamma c_1/\gamma_n < 2$. Combining this with (16) shows that this implies $n > 3$, so the above gives all cases where $N_+ = 1$ and $N > 3$.

Below I tabulate all the different versions that I have noted, along with the first known (to me) dates and sources, the numbers of solutions, given as (N, N_+), and the number of solutions given in the source. Tropfke [20] and Glaisher [12] cite a number of sources that I have not seen.

The Indian Version As mentioned earlier, the Indian version is somewhat different and gives infinitely many solutions. To illustrate, consider the following problem from Bhaskara, used as an example in Ayyangar.

> Example instanced by ancient authors: a stanza and a half. Three traders, having six, eight, and a hundred, for their capitals respectively, bought leaves of betle [or fruit] at an uniform rate; and resold [a part] so: and disposed of the remainder at one for five panas; and thus became equally rich. What was [the rate of] their purchase? and what was [that of] their sale?

This requires some explanation. Here the numbers c_i are not the numbers of objects, but the capitals (in panas) of each trader. It is assumed that you can buy D items per pana, so the the numbers of items will be Dc_i and our equation (1) becomes

$$a_i + b_i = Dc_i \,. \tag{1'}$$

Further, we are specifically given the greater price $A = 5$, and it is understood that the lesser price corresponds to selling β items per pana, i.e.,

Version	Date	Source	# of Solutions	Solutions in Source
(10, 30)	1202?	Fibonacci	(55, 36)	5 given.
(12, 32)	1202?	Fibonacci	(78, 55)	6 given.
(12, 33)	1202?	Fibonacci	(78, 55)	1 given.
(10, 20)	c. 1300	"Abraham"	(55, 36)	
(10, 30, 50)	1300s	Munich codex 14684	(25, 16)	1 given.
(30, 56, 82)	1489	Widman	(225, 196)	1 given.
(17, 68, 119, 170)	1489	Widman	(45, 35)	1 given.
(305, 454, 603, 752, 901)	1489	Widman	(11,552; 11,400)	1 given.
(10, 20, 30)	c. 1500	Pacioli	(25, 16)	1 given.
(8, 17, 26)	1513	Blasius	(16, 9)	1 given.
(20, 40, 60)	1515	Tagliente	(100, 81)	1 given.
(10, 50)	1521?	Ghaligai	(55, 36)	1 given.
(11, 33, 55)	1556	Tartaglia	(30, 20)	1 given.
(16, 48, 80)	1556	Tartaglia	(64, 49)	1 given.
(10, 20, . . . , 90)	1556	Tartaglia	(3, 1)	1 given.
(20, 30, 40)	1612	Bachet	(100, 81)	4 given.
(10, 25, 30)	1725?	Ozanam from De Lagny	(10, 6)	all given.
(18, 40)	1874	Labosne	(171, 136)	2 given, one with fractions.
(18, 40, 50)	1874	Labosne	(3, 1)	1 given with fractions.
(10, 12, 15)	1874	Labosne	(7, 4)	none given.
(31, 32, 37)	1874	Labosne	(70, 60)	none given.
(27, 29, 33)	1893	Hoffmann	(117, 100)	1 given.
(20, 30, . . . , 60)	1905	Dudeney	(45, 36)	1 given.
(20, 40, . . . , 140)	1924	Glaisher	(27, 21)	1 given.

$B = 1/\beta$, where β is an integer. Ayyangar's translation of the problem makes this more specific by saying they 'sold a part in lots and disposed of the remainder at one for five panas.' Thus (2) becomes

$$5a_i + \frac{b_i}{\beta} = C\,. \tag{2'}$$

We now have $2n+3$ unknowns and $2n$ equations, so we again have a three-dimensional solution set. One can scale the set D, a_i, b_i, C in some way to reduce the solution set to two dimensions, though it is not as easy to see how to do this as in the Western version. At this point, I must point out that some Indian versions give c_i and/or A as fractions! The Indian authors do not get very general solutions of these problems.

Sridhara finds a one-parameter solution set. Bhaskara makes the following comment: "This, which is instanced by ancient writers as an example of a solution resting on unconfirmed ground, has been by some means reduced to equation; and such a supposition introduced, as has brought out a result in an unrestricted case as in a restricted one. In the like suppositions, when the operation, owing to restriction, disappoints; the answer must by the intelligent be elicited by the exercise of ingenuity." Amen!

For ease of expression, we let $\beta_i = b_i/\beta$, so our equations become

$$a_i + \beta\beta_i = Dc_i\,; \tag{1''}$$

$$Aa_i + \beta_i = C\,. \tag{2''}$$

Ayyangar gives a reasonable solution process for this problem, but since his notation is different, I will sketch the solution in the present notation, pointing out an improvement. Eliminating β_i, then subtracting to eliminate C and using the same abbreviations as before, we get

$$D\beta_i = (A\beta - 1)\,\alpha_i \tag{4'}$$

and the condition for all the equations to have integral solutions is

$$A\beta - 1 \quad \text{divides} \quad D\Gamma\,. \tag{5'}$$

If we set

$$D\Gamma = k(A\beta - 1)\,, \tag{5''}$$

then (4′) gives

$$\alpha_i = \frac{k\gamma_i}{\Gamma}\,. \tag{6'}$$

We need to determine which k and β will make (5″) hold, i.e.,

$$k(A\beta - 1) \equiv 0 \pmod{\Gamma}\,. \tag{5'''}$$

Let $\gamma = (k, \Gamma)$. Then (5‴) is equivalent to $A\beta - 1 \equiv 0 \pmod{\Gamma/\gamma}$ and this is solvable if and only if $(A, \Gamma/\gamma) = 1$. This gives us a rather peculiar condition. Write $\Gamma = rs$, where r comprises all the prime powers in Γ whose primes also occur in A. Then Γ/γ is relatively prime to A if and only if Γ/γ divides s, which is if and only if r divides $\gamma = (k, \Gamma)$, which is if and only if r divides k. (Ayyangar notes only that (A, Γ) must divide k, and (A, Γ) can be a proper divisor of r.)

Both k and β can take on an infinite range of values, but in order for $\alpha_1 = k\gamma_n/\Gamma \leqslant Dc_1$, we must have

$$\beta \geqslant \frac{c_n}{Ac_1}\,. \tag{7'}$$

For given k and β satisfying ($5'''$) and ($7'$), there are only a finite number of solutions. Analysis similar to that done before shows that there are then $1 + [k(A\beta c_1 - c_n)/\Gamma\beta]$ solutions.

By the way, Bhaskara solves his example by taking $k = 2$, so $k = \Gamma$ and $\beta = 110$, so $D = 549$, giving total numbers Dc_i of 3295; 4392; 54,900 and a_i of 3294; 3292; 3200 and $C = 16{,}470$ [4].

References

[1] Alcuin. *Problems to Sharpen the Young.* Translated by John Hadley; annotated by David Singmaster and John Hadley. *Math. Gaz.* **76,**(475), Mar 1992, pp. 102–126. See No. 16: "De duobus hominibus boves ducentibus — Two men leading oxen."

[2] Ayyangar, A. A. Krishnaswami. "A classical Indian puzzle-problem." *J. Indian Math. Soc.* **15,** 1923–24, pp. 214–223.

[3] Bachet, Claude-Gaspar. *Problèmes plaisants & délectables qui se font par les nombres.* 1st ed., P. Rigaud, Lyon, 1612; Prob. 21, pp. 106–115.
2nd ed., P. Rigaud, Lyon, 1624; Prob. 24, pp. 178–186..
Revised by A. Labosne, Gauthier-Villars, Paris. 3rd ed., 1874; 4th ed., 1879; 5th ed., 1884. 5th ed. reprinted by Blanchard, Paris, 1959; Prob. 24, pp. 122–126.

[4] Bhaskara, Bijanganita, aka Bhâskara, Bîjaganita, 1150. In: Henry Thomas Colebrooke, trans.; *Algebra, with Arithmetic and Mensuration from the Sanscrit of Brahmegupta and Bháscara.* John Murray, London, 1817. (There have been several reprints, including Sändig, Wiesbaden, 1973.) Chap. 6, v. 170, pp. 242–244.

[5] Bonnycastle, John. *An Introduction to Algebra, with Notes and Observations; designed for the Use of Schools, and Other Places of Public Education,* 1782. The first nine editions appeared "without any material alterations." In 1815, he produced a 10th ed., "an entire revision of the work," which "may be considered as a concise abridgment" of his two-volume *Treatise on Algebra,* 1813. I examined the 7th edition, J. Johnson, London, 1805, and the 13th edition, J. Nunn et al., London, 1824, which may be the same as the 1815 edition.

[6] Bunt, Lucas N. H., et al. *The Historical Roots of Elementary Mathematics.* Prentice-Hall, 1976, p. 33.

[7] Chiu Chang Suan Ching, "Nine Chapters on the Mathematical Art," c. 150 B.C.; Translated into German by K. Vogel; Neun Bücher arithmetischer Technik; Vieweg, Braunschweig, 1968. Nos. 10, 12, 13, pp. 86–88.

[8] Diophantos, *Arithmetica.* In: T. L. Heath; Diophantos of Alexandria; 2nd ed., Oxford University Press, 1910; reprinted by Dover, 1964. Book I, nos. 15, 18, 19, pp. 134–136.

[9] *The Demaundes Joyous.* Wynken de Worde, London, 1511. Facsimile with transcription and commentary by John Wardroper, Gordon Fraser Gallery, London, 1971, reprinted 1976. This is the oldest riddle collection printed in England, surviving in a single example in the Cambridge University Library. Prob. 50, p. 6 of the facsimile, pp. 26–27 of the transcription.

[10] Euclid, *Opera.* Edited by J. L. Heiberg and H. Menge, Teubner, Leipzig, 1916. Vol. VIII, pp. 286–287.

[11] Fibonacci, aka Leonardo Pisano. *Liber Abbaci.* (1202); 2nd edition, 1228. In: *Scritti di Leonardo Pisano,* vol. I, edited and published by B. Boncompagni, Rome, 1857, pp. 298-302.

[12] Glaisher, J. W. L. "On certain puzzle-questions occurring in early arithmetical writings and the general partition problems with which they are connected." *Messenger of Mathematics,* **53,** 1923-24, pp. 1-131.

[13] Alkarkh, Aboû Beqr Mohammed Ben Alhaçen, aka al-Karagi. Untitled manuscript called "Alfakhrî," c. 1010. MS 952, Supp. Arabe de la Bibliothèque Impériale, Paris. Edited into French by Franz Woepcke as "Extrait du Fakhrî," L'Imprimerie Impériale, Paris, 1853. Reprinted by Georg Olms Verlag, Hildesheim, 1982. Sect. 3, no. 5, p. 90.

[14] Loyd, Sam. *Sam Loyd's Cyclopedia of 5,000 Puzzles, Tricks and Conundrums.* Edited by Sam Loyd II. Lamb Publishing, 1914; Pinnacle or Corwin, 1976, pp. 53 and 346.

[15] Mahavira, aka Mahâvîrâ(cârya). "Ganita-sâra-sangraha," A.D. 850. Translated by M. Ragacarya. Government Press, Madras, 1912.

[16] Metrodorus (compiler). *The Greek Anthology.* Translated by W. R. Paton. Loeb Classical Library, Harvard University Press, Cambridge, Mass., and Heinemann, London, 1916-18., Vol. 5.

[17] Ozanam, Jacques. *Récréations Mathématiques et Physiques,* Paris, 1694 (not seen). Reprint, Amsterdam, 1696; prob. 24, pp. 79-80. English version: *Recreations Mathematical and Physical,* R. Bonwick, et al., London, 1708; prob. 24, pp. 68-70. New edition, edited by Grandin, four vols., C. A. Jombert, Paris, 1725; prob. 28, pp. 201-210.

[18] Sridhara, aka Śrîdharâcârya, Pâtîganita, c. 900. Transcribed and translated by K. S. Shukla. Lucknow University, Lucknow, 1959. V. 60-62, ex. 76-77, pp. 44-49 and 94.

[19] Tartaglia, Nicolo. *General Trattato di Numeri et Misure.* Curtio Troiano, Venice, 1556. Part 1, book 16, prob. 136-139, pp. 256r-256v.

[20] Tropfke, Johannes, *Geschichte der Elementarmathematik.* Revised by Kurt Vogel, Karin Reich, and Helmuth Gericke. 4th edition, Vol. 1: Arithmetik und Algebra. De Gruyter, Berlin, 1980.

Who Wins Misère Hex?

Jeffrey Lagarias and Daniel Sleator

Hex is an elegant and fun game that was first popularized by Martin Gardner [4]. The game was invented by Piet Hein in 1942 and was rediscovered by John Nash at Princeton in 1948.

Two players alternate placing white and black stones onto the hexagons of an $N \times N$ rhombus-shaped board. A hexagon may contain at most one stone.

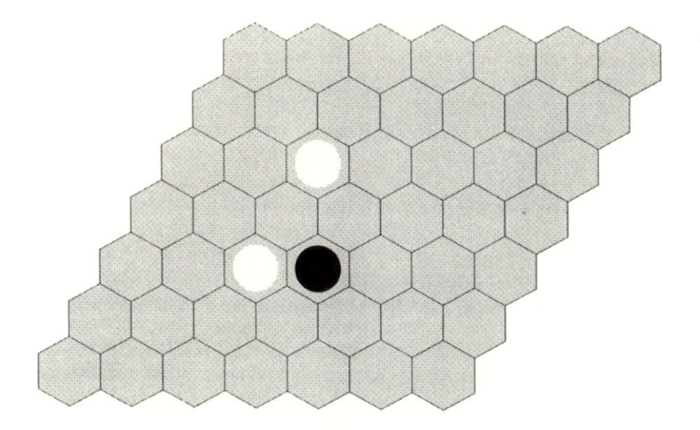

A game of 7×7 Hex after three moves.

White's goal is to put white stones in a set of hexagons that connect the top and bottom of the rhombus, and Black's goal is to put black stones in a set of hexagons that connect the left and right sides of the rhombus. Gardner credits Nash with the observation that there exists a winning strategy for the first player in a game of hex.

The proof goes as follows. First we observe that the game cannot end in a draw, for in any Hex board filled with white and black stones there must be either a winning path for white, or a winning path for black [1, 3]. (This fact is equivalent to a version of the Brouwer fixed point theorem, as shown by Gale [3].) Since the game is finite, there must be a winning strategy for either the first or the second player. Assume, for the sake of contradiction, that the second player has a winning strategy. The first player can make an arbitrary first move, then follow the winning strategy (reflected) for a second player (imagining that the hexagon containing the first move is empty). If the strategy requires the first player to move in this non-empty cell, the player simply chooses another empty cell in which to play, and now imagines that this one is empty. Since the extra stone can *only help* the first player, the winning strategy will work, and the first player wins. This contradicts our assumption that the second player has a winning strategy. Of course this proof is non-constructive, and an explicit winning strategy for the first player is not known.

The purpose of this note is to analyze a variant of Hex that we call *Misère Hex*. The difference between normal Hex and Misère Hex is that the outcome of the game is reversed: White wins if there is a black chain from left to right, and Black wins if there is a white chain from top to bottom. Misère Hex has also been called Reverse Hex and Rex.

Contrary to one's intuition, it is *not* the case that the second player can always win at Misère Hex. In fact, the winner depends on the parity of N; on even boards the first player can win, and on odd boards the second player can win.

This fact is mentioned in Gardner's July 1957 column on Hex. Gardner attributes the discovery to Robert Winder, who never published his proof. As in the case of Hex, the proof of the existance of a winning strategy does not shed any light on what that strategy is. A small step was made in that direction by Ron Evans [2] who showed that for even N, the first player can win by moving in an acute corner. An abstract theory of "Division games," which includes Hex and Misère Hex as special cases, was later developed by Yamasaki [5].

Here we present an elementary proof showing who wins Misère Hex. In addition to showing who wins, our result shows that in optimal play the loser can force the entire board to be filled before the game ends.

Theorem: *The first player has a winning strategy for Misère Hex on an $N \times N$ board when N is even, and the second player has a winning strategy when N is odd. Furthermore, the losing player has a strategy that guarantees that every cell of the board must be played before the game ends.*

Proof: It suffices to prove the second assertion, because it shows that the parity of the number of cells on the board determines which player has the winning strategy.

Because the game cannot end in a draw, either the second player or the first player has a winning strategy. Let P be the player who has a winning strategy, and let Q be the other player. For any winning strategy $\mathcal{L}$ for P define $m(\mathcal{L})$ to be the minimum (over all possible games of Misère Hex in which P plays strategy $\mathcal{L}$) of the number of cells left uncovered at the end of the game. We must show that $m(\mathcal{L}) = 0$.

We shall make use of the following *monotonicity property* of the game. Consider a terminal position of a game that is a win for Q. By definition, such a position contains a P-path. Suppose the position is modified by filling in any subset of the empty cells with Q's stones, and further modified by changing any subset of Q's stones into P's stones. The position is still a win for Q, because none of these changes would interfere with the P-path.

We are now ready to prove the theorem. We shall argue by contradiction, supposing that $m(\mathcal{L}) \geq 1$. The contradiction will be to show that under this assumption Q has a winning strategy. The basic idea resembles Nash's proof that the first player has a winning strategy for Hex, in that we will describe a new strategy for Q in which (in effect) Q makes an extra move and then plays the reflected version $\mathcal{L}^R$ of P's hypothetical winning strategy $\mathcal{L}$. (Note that $m(\mathcal{L}) = m(\mathcal{L}^R)$.) The proof is complicated, however, by the fact that it is not clear *a priori* that having an extra stone on the board is either an advantage or a disadvantage. The proof splits into two cases depending on whether Q is the first player or the second.

Suppose that Q is the first player. Player Q applies the following strategy. She makes an arbitrary first move, and draws a circle around the cell containing this move. From now on she applies strategy $\mathcal{L}^R$ in what we shall call the *imaginary game*. The state of this game is exactly like that of the real game, except that in the imaginary game the encircled cell is empty, while in the real game, that cell contains a Q-stone. This relationship will be maintained throughout the game. When the strategy $\mathcal{L}^R$ requires Q to play in the encircled cell, she plays instead into another empty cell (chosen arbitrarily), erases the circle, and draws a new circle around the move just played. Because $m(\mathcal{L}^R) \geq 1$, when it is P's turn to move there must be at least two empty cells in the imaginary game, and there must be at least one empty cell in the real game. Therefore it is possible for P to move. (We'll see below that P will not have won the real game.) Similarly, when it is Q's turn to move there must be at least three empty cells in the imaginary game, so there are at least two empty cells in the real game. Thus the real game can continue.

Eventually Q will win the imaginary game because $\mathcal{L}^R$ is a winning strategy. When this happens she has also won the real game, because of the monotonicity property. This contradicts our assumption that P has a winning strategy.

Now, suppose that Q is the second player. Let p_0 be P's first move. Player Q begins by encircling p_0, playing out $\mathcal{L}^R$ in an imaginary game. The imaginary game and the real game differ in up to two places, as follows. The imaginary game is obtained from the real game by first changing p_0

from P's stone to Q's stone, then by erasing the stone in the encircled cell. If the strategy $\mathcal{L}^R$ requires a move into the encircled cell, then Q arbitrarily chooses a different empty cell in which to move, and transfers the circle from its current location to the new cell. The fact that $m(\mathcal{L}^R) \geq 1$ ensures that both players can continue to move. It is easy to see that the relationship between the real game and the imaginary game is maintained.

Player Q eventually wins the imaginary game. The position in the real game is obtained from the position in the imaginary game by putting Q's stone in the encircled cell, and changing the contents of p_0 from a Q-stone to a P-stone. The position in the imaginary game is a winning position for Q, and the monotonicity property ensures that the corresponding position in the real game is also a win for Q. This contradicts our assumption that P has a winning strategy. □

References

[1] A. Beck, M. Bleicher, and J. Crow, *Excursions into Mathematics,* Worth, New York, 1969, pp. 327–339.

[2] R. Evans, A winning Opening in Reverse Hex, *J. Recreational Mathematics,* **7**(3), Summer 1974, pp 189–192.

[3] D. Gale, The Game of Hex and the Brouwer Fixed-Point Theorem, *The American Mathematical Monthly,* **86** (10), 1979, pp. 818–827.

[4] M. Gardner, *The Scientific American Book of Mathematical Puzzles and Diversions,* Simon and Schuster, New York, 1959, pp. 73–83.

[5] Y. Yamasaki, Theory of Division Games, *Publ. Res. Inst. Math. Sci., Kyoto Univ.* **14**, 1978, pp. 337–358.

An Update on Odd Neighbors and Odd Neighborhoods

Leslie E. Shader

Some time ago (1986–88) this problem appeared in mathematical circles:

Problem 1: Can the unit squares of an $n \times n$ sheet of graph paper be labeled with 0's and 1's so that every neighborhood is odd?

A *neighborhood*, N_i, of square i, is the set of squares sharing an edge with square i and does include square i. A neighborhood, N_i is *odd* if there are an odd number of squares in N_i with labels "1".

Figure 1 shows a 4×4 example with a desired labeling.

1	1	0	1
1	1	1	0
0	1	1	1
1	0	1	1

Figure 1.

The first solution of Problem 1 was said to be quite difficult. Later, cellular automata ideas were used. However, if the problem is generalized to all graphs the proof is quite elementary.

Theorem 1. *Let G be a graph with no loops or multiple edges. Let A be the adjacency matrix of G. Then the vertices of G can be labeled with 0's and 1's so that every neighborhood is odd.*

Proof:[1] Consider the matrix equation $(A+I)x = 1$ where 1 is a column of 1's. We need to show that this equation has a solution for every adjacency matrix A. The linear algebra is, of course, over $\mathbf{Z}_2$, the integers modulo 2, since we are interested in only odd/even properties. Suppose y is a $1 \times n$ vector of 0's and 1's. Then

$$y(A+I) = 0 \iff yA = yI \iff Ay^T = yIy^T = yy^T.$$

[1]Due to the author's son, Bryan, while a graduate student.

But A is symmetric, so $yAy^T = 0$. Therefore, $yy^T = 0$ and rank$(A+I) =$ rank $(A + I|1)$, so $(A + I)x = 1$ for every adjacency matrix A. Finally, for every graph G, the vertices of G can be labeled with 0's and 1's so that every neighborhood in G is odd.

If we think of the unit squares in Problem 1 as vertices of G and v_iv_j is an edge in $G \iff v_i$ shares an edge with v_j, then Theorem 1 implies that the squares of the $n \times n$ square can be labeled with 0's and 1's so that each neighborhood is odd.

Now for some new questions (at least at the time that this paper was submitted).

The Developer's Dilemma

Suppose a real estate developer has an infinite (at least sufficiently long) strip of land m lots wide.

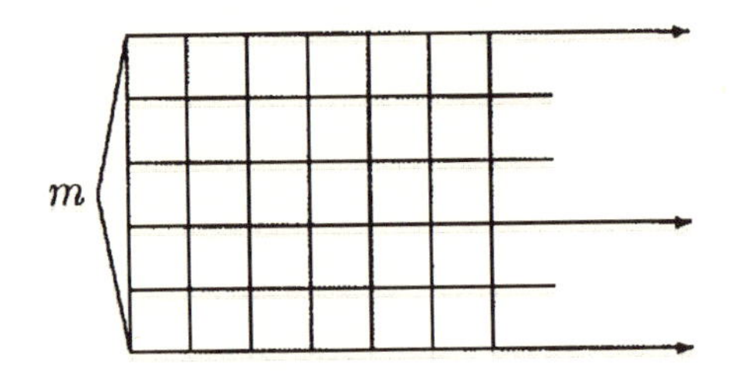

Figure 2. $m = 5$.

Further, we assume all lots in the first strip of m lots have been sold. It is quite possible that some of the buyers are odd, and some are not odd. Is there an n so that the developer can sell all the lots in the $m \times n$ rectangle and satisfy the politically correct criteria that all neighborhoods have an odd number of odd neighbors?

When this paper was originally written, the answer to the Developer's Dilemma was unknown. We will generalize, as before, to all graphs and prove a theorem. But this time the generalization does not yield a solution to the Developer's Dilemma. However, a proof that the Dilemma can be solved will also be presented.

Theorem 2. *Let G be a graph with vertices labeled with 0's and 1's in any way. If not all neighborhoods of G are odd, then G can be embedded in a graph G', with all neighborhoods in G' odd.*

It is rather interesting that only one vertex must be added to G in order to prove Theorem 2.

Proof. Let G' be the graph formed from G by adding a new vertex, w. Edges from w to vertices $v \in G$, with N_v even, are also added. Label w with a 1. Clearly all neighborhoods N_v with $v \neq g, v \in G'$ are now odd in G'. We need only to show N_w is odd in G'. Let D_k be the number of vertices labeled 1 and adjacent to vertex $k \in G$. N_k is even in $G \iff D_k$ is odd and N_k is odd in $G \iff D_k$ is even. But,

$$\sum_{\substack{k \in G \\ k \text{ labeled } 1}} D_k \text{ is even,}$$

since the sum counts the edges from vertices labeled 1 exactly twice. Hence, there are an even number of vertices $k \in G$ labeled 1 and N_k is even in G. These vertices now are all connected to w with label 1. So N_w is odd $\in G'$.

Theorem 3. *The Developer's Dilemma can be solved.*

Proof. Let the $m \times \infty$ strip of unit squares be horizontal. We call v_1, any 0–1 m-tuple, a "starter" for the $m \times n$ rectangle we seek. We wish to construct a proper labeling of an $m \times n$ rectangle so that all neighborhoods are odd and the vector v_1 is the "starter" column.

Perhaps the $m \times 1$ rectangle using the starter v_1 has the property that all neighborhoods are odd. Figure 3 shows that this phenomenon can happen.

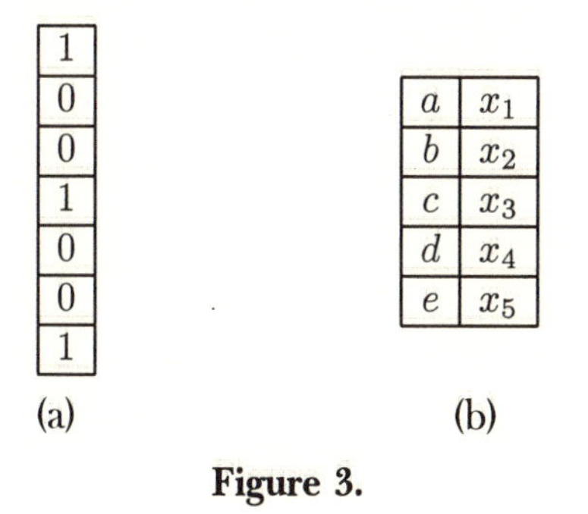

Figure 3.

If not, all neighborhoods are odd, and we describe an algorithm that can be used to label the next column so that all N_{1k} are odd for v_{1k} and unit square in column 1. (See Figure 3(b).)

If $a+b$ is odd, let $x_1 = 0$, otherwise $x_1 = 1$.
If $a+b+c$ is odd, let $x_2 = 0$, otherwise $x_2 = 1$.
If $b+c+d$ is odd, let $x_3 = 0$, otherwise $x_3 = 1$.
If $c+d+e$ is odd, let $x_4 = 0$, otherwise $x_4 = 1$.
If $d+e$ is odd, let $x_5 = 0$, otherwise $x_5 = 1$.

Of course, m is not limited to the value 5. Now that every neighborhood for v_{1k} is odd, we can repeat the process for column 2. Each neighborhood will now have all but one entry labeled, and that label can be assigned so that the neighborhood is odd.

This process can be repeated as many times as we like, and each new column will be dependent on just the preceding two. We will have a solution to our problem if the $n+1$ column has all entries 0, i.e., N_{nk} is odd for square v_{ik} in column $n, i \leq n$. Since there are only a finite number of adjacent pairs of the m-tuple, either we get a column of 0's or there must be an adjacent pair that repeats in our process. Let $v_i v_{i+1}$ be the first pair that repeats. We can generate new columns moving from left to right or from right to left, and we generate a second $v_i v_{i+1}$ somewhere to the right of the first $v_i v_{i+1}$ (see Figure 4).

$$\begin{array}{ccccccc} v_1 & \cdots & v_i & v_{i+1} & \cdots & v_i & v_{i+1} \\ \downarrow & & \downarrow & \downarrow & & \downarrow & \downarrow \end{array}$$

Figure 4.

Using the second occurrence of $v_i v_{i+1}$, column v_{i-1} must repeat to the left of the first occurrence of $v_i v_{i+1}$.

$$v_1 \cdots v_{i-1} v_i v_{i+1} \cdots v_{i-1} v_i v_{i+1}$$

So $v_i v_{i+1}$ is not the first repeating pair. Contradiction! Hence there is a column of 0's and the proof is complete.

Figure 5 shows an example where $v_1 = \mathrm{col}(00110)$, and $n = 7$.

0	1	0	1	0	0	0	0
0	0	1	0	0	1	1	0
1	1	0	1	0	1	1	0
1	1	0	0	1	0	0	0
0	0	0	1	0	1	0	0

Figure 5.

Perhaps the next question might be *For m fixed, is there an n such that for every starter v the $m \times n$ rectangle has all neighborhoods odd?, i.e., n is independent of v.*

For $m = 5$, to show that $n = 23$ is a nice exercise for the reader.

Point Mirror Reflection

M. Oskar van Deventer

The Problem

It is well known that a ray of light can reflect many times between two ordinary line mirrors. If we introduce the condition that a ray should reflect only one point on the mirror – reducing the mirrors to point mirrors – we find a maximum of three reflections for two mirrors and seven reflections for three mirrors, if these are suitably placed and oriented. Solutions are shown in Figure 1.

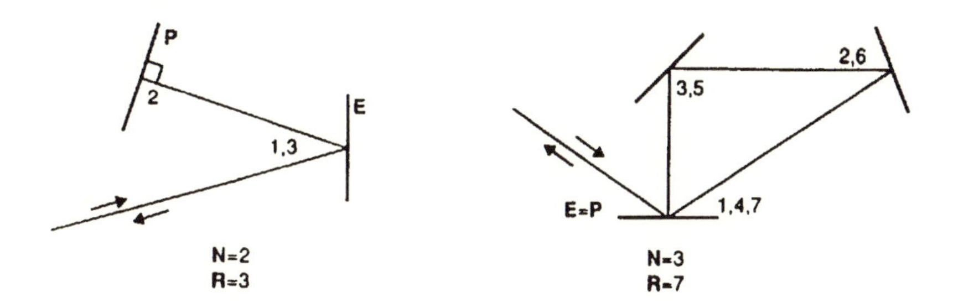

Figure 1. Maximum number of reflections for two and three point mirrors (E = entrance mirror, P = perpendicularly reflecting mirror).

This suggests an intriguing problem: How must N point mirrors be placed and oriented to reflect an entering ray of light as often as possible between these mirrors? What is the maximum number of reflections for varying N?

Upper Limits. From Figure 1 we can see that after some reflections the ray of light is reflected perpendicularly onto a mirror P and returns along the same paths but in the opposite direction. If the ray would not return along the entry paths, passing once again by entrance mirror E, half of the potential reflections would not be used. It is obvious that the strategy of

Figure 2. Number of reflections for different mirror types.

using a perpendicularly reflecting mirror P yields the maximum number of reflections.

At each mirror a maximum of $N-1$ rays can be reflected, going to every other mirror. At mirror E there can be one additional reflection for the entering ray. Therefore, the theoretical maximum number of reflections is $N(N-1)+1$, or

$$R_1 = N^2 - N + 1\,. \tag{1}$$

However, for even N there is a parity complication. Only at mirror P can there be an odd number of reflections. At all other mirrors there must be an even number of reflections, since the light paths are used in two directions; see Figure 2.

At mirror E there can be at most N reflections, $N-1$ from the other mirrors and one from the entering ray. At mirror P there can be at most $N-1$ reflections, from the other mirrors, which is an odd number. However, at the other $N-2$ mirrors (other than the E and P mirrors) there can't be $N-1$ reflections, but only $N-2$ to make it an even number. So the upper limit for an even number of mirrors is $N+(N-1)+(N-2)(N-2)$ or

$$R_2 = N^2 - 2N + 3\,. \tag{2}$$

For odd N the parities are okay if E and P are the same mirror, so R_1 remains valid.

Mirrors on the Corners of a Regular Polygon

Can the theoretical maxima R_1 and R_2 be achieved for any N, and what should the positions and the orientations of the mirrors be? A well-known geometrical property may help us: When the N mirrors are put on the corners of a regular N-gon, the angle between all the neighboring pairs of light paths is $180°/N$, as shown for the regular pentagon in Figure 3.

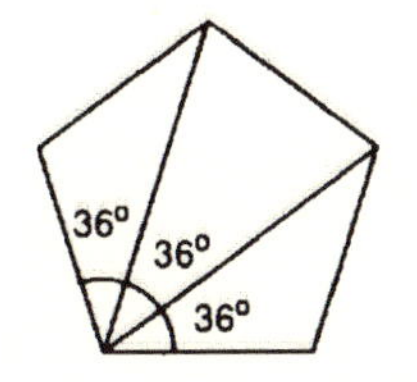

Figure 3. Light paths in a regular pentagon.

Using such an N-gon grid, two-dimensional space is sufficient to find solutions for each N. Only discrete orientations of the mirror are allowed and are multiples of $90°/N$, say $m \times 90°/N$. For $m = 0$ we have the normal orientation, $m = 1$ yields a rotation of one step clockwise, and so on. This is illustrated in Figure 4.

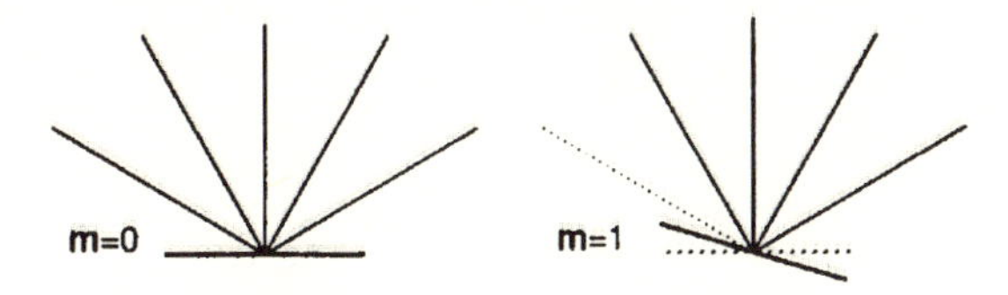

Figure 4. Rotation of a mirror.

Prime Number of Mirrors. For prime N the maximum number of reflections can always be achieved. We take $m = 1$ for E (= P) and $m = 0$ for all other mirrors. The ray of light, starting at mirror E, makes hops of one mirror; after passing mirror E again, it makes hops of two mirrors, and so on. Since N is prime, the ray will pass every mirror once at each tour from E to E; see Figure 5.

Non-Prime Number of Mirrors. If the number of mirrors is non-prime, the investigation of various cases is less straightforward. A systematical trial-and-error search for the maximum number of reflections may be performed using the following strategy:

Step 1. Draw the N-gon and all possible light paths (the N-gon grid). Add the starting ray at an angle of $180°/N$ with the N-gon.

Step 2. Omit one or more light paths (lines of the N-gon grid), and adjust the mirror orientations such that symmetry of light paths is preserved at every mirror and such that only at mirror P is there a perpendicular reflection. Prevent any obvious loops, since a ray can neither enter nor leave such a loop.

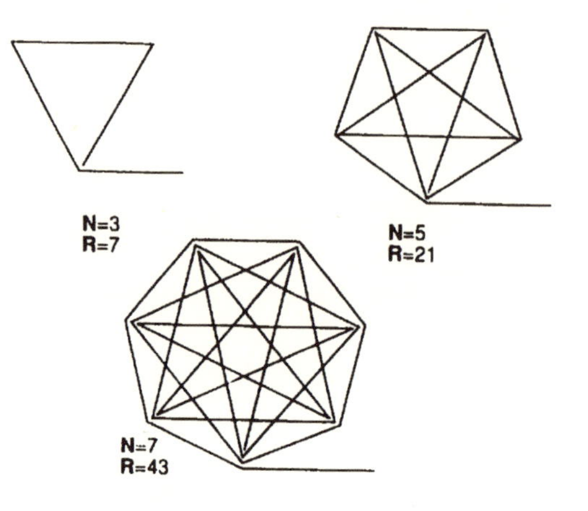

Figure 5. Reflection diagrams for prime numbers of mirrors.

Step 3. Count the number of reflections by following the entering ray. Calculate whether the number of reflections is equal to two times the number of remaining light paths plus one. If this is true, there are no loops and all of the remaining light paths are used. If not, reject the candidate solution.

Repeat Steps 2 and 3 for all candidates missing one light path, then for all candidates missing two light paths, and so on, until one or more solutions are found.

Figure 6 illustrates the procedure for the case of six mirrors. At least two light paths have to be omitted for Step 2. The candidate has 13 light paths. The grid consists of 15 lines, a choice of a set of 2 points out of 6, decreased by 2 to account for the loss of the two vertical lines on the left and right sides. In this case, however, we do not count $2 \times 13 + 1 = 27$ reflections but only 19. Indeed, there is a loop and the candidate has to be rejected.

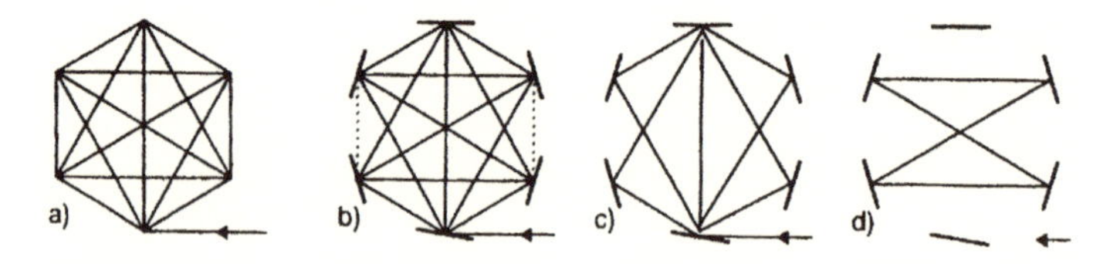

Figure 6. Searching strategy for $N = 6$; (a) Hexagon grid plus entering ray; (b) A candidate having 19 reflections and a loop, shown separately in (c) and (d).

N	R	d	m															
4	11	1	0	*0	0													
6	27	2	1	1	-1	1	-1	*0										
6	27	2	1	0	1	-1	0	*0										
8	51	3	1	1	-1	*0	1	-1	1	-1								
9	69	2	1	0	0	0	*1	-1	0	0	0							
10	83	4	1	1	-1	*0	0	1	-1	1	-1	0						
12	123	5	1	1	-1	1	-1	*0	1	-1	1	-1	1	-1				
14	171	6	1	1	-1	1	-1	1	-1	1	-1	1	-1	1	-1	*0		
15	199	6	*1	0	0	0	0	0	0	0	0	2	0	-2	1	-1	0	
16	227	7	1	1	-1	*0	0	0	0	0	1	-1	1	-1	0	0	0	0
		$c \rightarrow$	1	2	3	4	5	6	7	8	9	10	11	12	13	14	15	16

N = number of mirrors. c = corner number. m = mirror orientation number.
R = number of reflections. d = number of deleted paths.
The entrance mirror E is positioned in corner number one ($c_E = 1$).
The position of the perpendicular reflecting mirror P is marked with an asterisk*.

Figure 7. Data for composite N up to 16.

Some results obtained by applying the strategy outlined above are summarized in Figures 7 and 8.

- For N composite and odd the theoretical maximum of R_1 can never be achieved. If no light paths are omitted, m must be 0 for all mirrors, except for E, which has $m = 1$. There will always be an N/n loop, in which n stands for any divisor of N. For this case a sharper theoretical maximum might be defined, since light paths have to be omitted to account for all N/n loops.
- For even N this sharper maximum is given by R_2. For $N = 2$ to $N = 16$, the value R_2 can be achieved. Actual solutions for $N = 18$ or $N = 20$ have not been found, but might be possible.
- For $N = 6$ there are two solutions; see Figures 7, 8, and 9. For all other N up to 16 there is only one solution.

A general solution of the problem has not been found, nor a proof that a solution can always be put on our N-grid. It is likely that this is true, since only the N-gon grid seems to have a geometry that suits the problem.

Conclusion

A problem was defined regarding the maximum number of reflections of a ray of light between N point mirrors. A theoretical maximum was found,

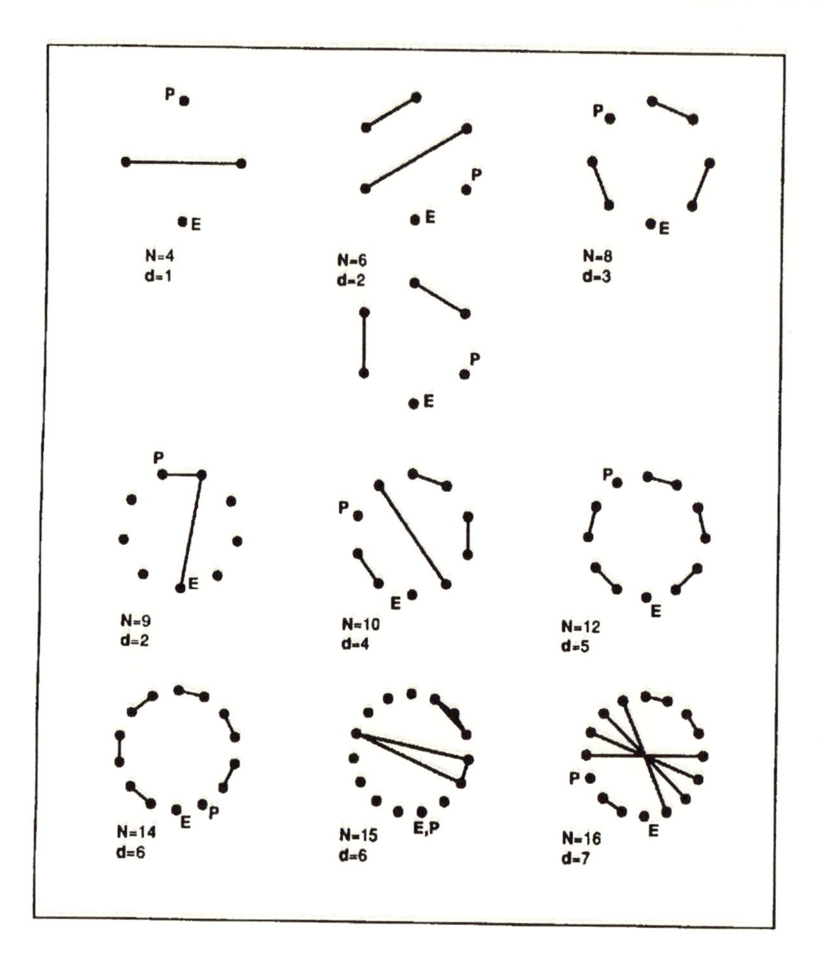

Figure 8. Deleted paths for composite N up to 16.

which can be reached for prime N. For composite N up to 16, solutions were found by a systematic trial and error search using a regular N-gon grid. No general strategy was found, and many questions remain open:

- Is there a better search strategy for composite N?
- What are the properties of solutions for $N = 18$ and higher?
- Are there more N with multiple solutions?
- Can a sharper theoretical maximum for odd and composite N be expressed in a closed form?
- Are there solutions that use the back side of the mirrors?

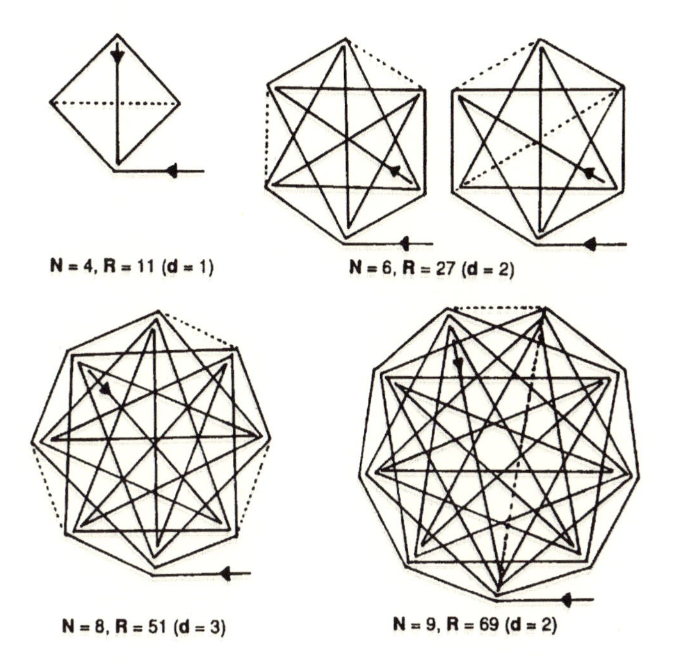

Figure 9. Reflection diagrams for composite N up to 9.

- Are there solutions that can't be put on an N-gon grid or that require three-dimensional space?
- Is there a general theory to the problem?
- Are there applications to, for example, lasers or burglar alarms?

How Random Are $3x + 1$ Function Iterates?

Jeffrey C. Lagarias

Abstract. The $3x + 1$ problem concerns the behavior under iteration of the $3x + 1$ function $T : \mathbb{Z} \to \mathbb{Z}$, defined by

$$T(n) = \begin{cases} \frac{3n+1}{2} & \text{if } n \text{ is odd,} \\ \frac{n}{2} & \text{if } n \text{ is even.} \end{cases}$$

The $3x+1$ Conjecture asserts that any positive integer n eventually reaches 1 under iteration by T. Simple probabilistic models appear to model the average behavior of the initial iterates of a "random" positive integer n. This paper surveys results concerning more sophisticated probability models, which predict the behavior of "extreme" trajectories of $3x + 1$ iterates. For example, the largest integer occurring in a trajectory starting from an integer n should be of size $n^{2(1+o(1))}$ as $n \to \infty$, and trajectories should be of length at most 41.677647... log n before reaching 1. The predictions of these models are consistent with empirical data for the $3x + 1$ function.

1. Introduction

The well-known $3x + 1$ problem concerns the behavior under iteration of the $3x + 1$ function $T : \mathbb{Z} \to \mathbb{Z}$, defined by

$$T(n) = \begin{cases} \frac{3n+1}{2} & \text{if } n \text{ is odd,} \\ \frac{n}{2} & \text{if } n \text{ is even.} \end{cases} \tag{1.1}$$

The *3x+1 Conjecture* asserts that for each positive integer n there is some iterate k with $T^{(k)}(n) = 1$, where $T^{(2)}(n) = T(T(n))$, etc. The $3x + 1$ Conjecture combines simplicity of statement with apparent intractability. Huge numbers of computer cycles have been expended studying it. In particular, it has now been verified for all $n < 2.702 \times 10^{16}$; see Oliveira e Silva [19].

The $3x+1$ problem appeared in Martin Gardner's "Mathematical Games" column in June 1972 [10]. Prior to that it circulated for many years by word of mouth. It is usually credited to Lothar Collatz, who studied problems similar to it in the 1930s, and who has stated that he circulated the problem in the early 1950s [6]. It was also independently discovered by B. Thwaites in 1952; see [22]. The first mathematical papers on it appeared around 1976 ([8], [21]), and now more than one hundred papers have been written about the problem. Surveys of the known results on the $3x+1$ problem can be found in Lagarias [13], Müller [18], and Wirsching [25].

To describe iterates of the $3x+1$ function we introduce some terminology. The *trajectory* or *forward orbit* $\mathcal{O}(n)$ of the positive integer n is the sequence of its iterates

$$\mathcal{O}(n) := \left(n, T(n), T^{(2)}(n), T^{(3)}(n), \ldots\right). \tag{1.2}$$

Thus $\mathcal{O}(27) = (27, 41, 62, 31, 47, \ldots)$. The *total stopping time* $\sigma_\infty(n)$ of n is the minimal number of iterates k needed to reach 1, i.e.,

$$\sigma_\infty(n) := \min\left\{k : \; T^{(k)}(n) = 1\right\}, \tag{1.3}$$

where $\sigma_\infty(n) = \infty$ if no iterate equals 1. We let $t(n)$ denote the largest value reached in the trajectory of n, so that

$$t(n) := \sup\left\{T^{(k)}(n) \; : \; k \geq 0\right\}, \tag{1.4}$$

where $t(n) = \infty$ if the sequence of iterates is unbounded. Finally we let $r(n)$ count the fraction of odd iterates in a trajectory up to the point that 1 is reached, i.e.,

$$r(n) := \frac{1}{\sigma_\infty(n)} \# \left\{k \; : T^{(k)}(n) \equiv 1 \pmod{2} \text{ for } 0 \leq k \leq \sigma_\infty(n)\right\}. \tag{1.5}$$

The $3x+1$ Conjecture is made plausible by the observation that $3x+1$ iterates behave "randomly" in some average sense. This randomness property can be formalized by taking an input probability distribution on the integers and looking at the probability distribution resulting from iterating the $3x+1$ function some number of times. For example, if one takes the uniform distribution on $[1, 2^k]$, then the distribution of the k-vector

$$v_k(n) := \left(n, T(n), \ldots, T^{(k-1)}(n)\right) \pmod{2}$$

is exactly uniform, i.e., each possible binary pattern occurs exactly once in $1 \leq n \leq 2^k$. Furthermore the resulting pattern is periodic in n with period

2^k. This basic property was independently discovered by Everett [8] and Terras [21].

One may summarize this by saying that the parity of successive iterates of T initially behaves like independent coin flips. Since a number n is multiplied by $\frac{1}{2}$ if it is even and approximately $\frac{3}{2}$ if it is odd, one expects that on average it changes multiplicatively by their geometric mean, which is $(\frac{3}{4})^{1/2}$. Since this is less than 1, one expects the iterates to decrease in size and eventually become periodic.

Several heuristic probabilistic models describing $3x+1$ iterates are based on this idea; see [9], [13], [15], [20], [24]. The simplest of them assumes that this independent coin-flip behavior persists until 1 is reached under iteration. These mathematical models predict that the expected size of a "random" n after k steps should be about $(\frac{3}{4})^{k/2}n$, so that the average number of steps to reach 1 should be

$$(1.6) \qquad \left(\frac{1}{2}\log\frac{4}{3}\right)^{-1} \log n \approx 6.95212 \log n \ .$$

Furthermore, the number of steps $\sigma_\infty(n)$ for n to iterate to 1 should be normally distributed with mean $\left(\frac{1}{2}\log\frac{4}{3}\right)^{-1}\log n$ and variance $c_1\sqrt{\log n}$, for an explicit constant c_1; see [20], [24]. There is excellent numerical agreement of this model's prediction with $3x+1$ function data.

The pseudorandom character of $3x+1$ function iterates can be viewed as the source of the difficulty of obtaining a rigorous proof of the $3x+1$ Conjecture. On the positive side, Cloney, et al. [5] proposed using the $3x+1$ function as a pseudorandom number generator. A well-developed theory of pseudorandom number generators shows that even a single bit of pseudorandomness can be inflated into an arbitrarily efficient pseudorandom number generator; cf. Lagarias [14] and Luby [17]. Even though $3x+1$ function iterates possess quite a bit of structure (cf. Garner [11] and Korec [12]) they also seem to possess some residual structurelessness, which may be enough for a pseudorandom bit to be extracted.

From this viewpoint it becomes interesting to determine how well the properties of $3x+1$ iterates can be described by a stochastic model. In doing so we are in the atypical situation of modeling a purely deterministic process with a probabilistic model. Here we survey some recent results on stochastic models, obtained in joint work with Alan Weiss and David Applegate, which concern extreme behaviors of $3x+1$ iterates.

Lagarias and Weiss [15] studied two different stochastic models for the behavior of $3x+1$ function iterates. These models consist of a repeated random walk model for forward iterates by T, and a branching random walk model for backward iterates of T. The repeated random walk model results of [15] show almost perfect agreement between empirical data for $3x + 1$ function iterates for n up to 10^{11}. This model has the drawback that it does not model the fact that $3x+1$ trajectories are not independent; indeed actual $3x+1$ trajectories coalesce to form a tree structure. The local structure of backward $3x+1$ iterates can be explicitly described using $3x+1$ trees, which are defined in Section 3. Lagarias and Weiss introduced a family of branching random walk models to describe the ensemble of such trees. These models make a prediction for extreme values of $\sigma_\infty(n)$ that is shown to coincide with that made by the repeated random walk model. That is, for these probabilistic models the deviation from independence exhibited by $3x + 1$ trajectories does not affect the asymptotic maximal length of extremal trajectories.

More recently Applegate and Lagarias ([1] and [3]) studied properties of the ensemble of all $3x + 1$ trees. They compared empirical data with predictions made from stochastic models, and found small systematic deviations of the distribution of $3x+1$ inverse iterates from that predicted by the branching random walk model above. They observed that the distribution of the largest and smallest number of leaves possible in such trees of depth k appears to be narrower than what would be predicted by the model of Lagarias and Weiss. This leads to two conjectures stated in Section 4. It remains a challenge to exploit such regularities to obtain new rigorous results on the $3x + 1$ problem.

The $3x + 1$ Conjecture remains unsolved and is viewed as intractable. Various authors have obtained rigorous results in the direction of the $3x+1$ Conjecture, consisting of lower bounds for the number of integers n below a value x that have some iterate $T^{(k)}(n) = 1$. More generally, one may estimate for a positive integer a the quantity

$$\pi_a(x) := \text{card}\left\{n : 1 \le n \le x \text{ with some } T^{(k)}(n) = a\right\}.$$

It has been conjectured that for each positive $a \equiv 1$ or $2 \pmod 3$ there is a positive constant c_a such that

$$\pi_a(x) > c_a x \quad \text{for all} \quad x \ge a\,;$$

see Applegate and Lagarias [2] and Wirsching [25]. At present the best rigorous bound of this sort states that for positive $a \equiv 1$ or $2 \pmod 3$ one has

$$\pi_a(x) > x^{0.81}$$

for all sufficiently large x; see [2].

2. Repeated Random Walk Model

Lagarias and Weiss [15] studied a repeated random walk model for forward iterates by T. In this model, the successive iterates $\log T^{(k)}(n)$ are modeled by starting at $\log n$ and taking independent steps of size $-\log 2$ or $\log \frac{3}{2}$ with probability $\frac{1}{2}$ each. The random variable $\sigma_\infty^*(n)$ is the step number at which 0 is crossed. This random variable is finite with probability one, and the expected size of $\sigma_\infty^*(n)$ is given by (1.6). For each initial value n an entirely new random walk is used; this explains the name "repeated random walk model." What is the extremal size of $\sigma_\infty^*(n)$? Large deviation theory shows that with probability one

$$\limsup_{n\to\infty} \frac{\sigma_\infty^*(n)}{\log n} = \gamma_{RW} \simeq 41.677647 \ . \tag{2.1}$$

The constant γ_{RW} is the unique solution with $\gamma > (\frac{1}{2}\log\frac{4}{3})^{-1}$ of the functional equation

$$\gamma g\left(\frac{1}{\gamma}\right) = 1 \ , \tag{2.2}$$

where

$$g(a) := \sup_{\theta\in\mathbb{R}} \left[a\theta - \log\frac{1}{2}\left(2^\theta + \left(\frac{2}{3}\right)^\theta\right)\right] . \tag{2.3}$$

Next, let $t^*(n)$ equal the maximal value taken during the random walk, so that $t^*(n) \geq \log n$. Large deviation theory shows that, with probability 1,

$$\limsup_{n\to\infty} \frac{t^*(n)}{\log n} = 2 \ . \tag{2.4}$$

Finally, we consider the random variable

$$r^*(n) := \frac{1}{\sigma_\infty(n)} \# \left\{ k : T^{(k)}(n) > T^{(k-1)}(n) \text{ with } 1 \leq k \leq \sigma_\infty(n) \right\} .$$

Large deviation theory predicts that with probability 1 it satisfies

$$\limsup_{n\to\infty} r^*(n) = \rho \simeq 0.609090 \, .$$

The quantity ρ corresponds to an estimate of the maximum fraction of elements in a $3x+1$ trajectory that are odd. Large deviation theory also asserts that the graph of the logarithmically scaled trajectories

$$\left\{ \left(\frac{k}{\log n}, \frac{\log T^{(k)}(n)}{\log n} \right) : 0 \leq k \leq \sigma_\infty(n) \right\} \tag{2.5}$$

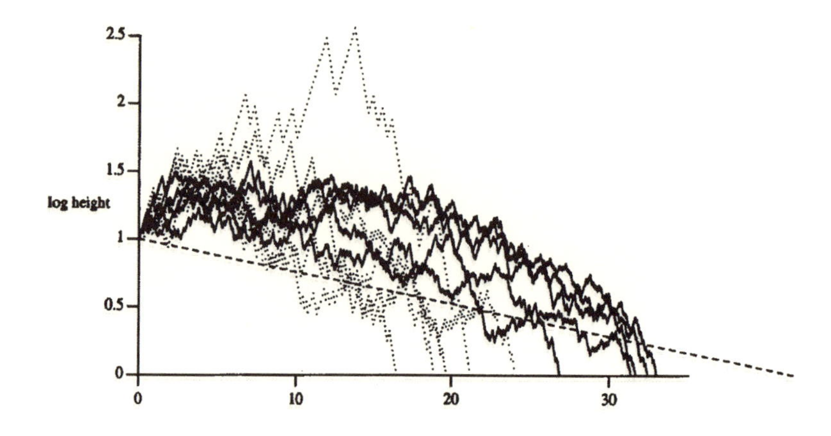

Figure 1. Scaled trajectories of n_k maximizing $\gamma(n)$ in $10^k \leq n \leq 10^{k+1}$ (dotted for $1 \leq k \leq 5$; solid for $6 \leq k \leq 10$).

approaches a characteristic shape. For the extremal trajectories for $\gamma^*(n)$ in (2.1) it is a straight line segment with slope -0.0231 starting from $(0, 1)$ and ending at $(\gamma, 0)$; see Figure 1. For the extremal trajectories for $t^*(n)$ in (2.4), the limiting graph of these trajectories has a different appearance. The length of such trajectories approaches the limiting value $\sigma_\infty(n) \asymp 21.55 \log n$, and the graph (2.5) approximates the union of two line segments, the first with slope 0.1318 for $1 \leq k \leq 7.645 \log n$, starting from $(0, 1)$, and ending at $(7.645, 2)$, and the second with slope -0.1439 for $7.645 \log n \leq k \leq 21.55 \log n$, starting at $(7.645, 2)$ and ending at $(21.55, 0)$; see Figure 2.

A comparison of (2.1) with numerical data up to 10^{11} is given in Table 1. It gives the longest trajectory, where $\gamma(n) = \frac{\sigma_\infty(n)}{\log n}$, and where $r(n)$ gives the fraction of odd entries in the iterates $(n, T(n), \ldots, T^{(k)}(n))$ in this trajectory.

The trajectories of these values of n are plotted in Figure 1, and the large deviations extremal trajectory is indicated by a dotted line. The agreement of the data up to 10^{11} with the repeated random walk model is quite good; an analysis in [15, Section 5] shows that in 10^{11} trials one should only expect to find a largest $\gamma(n) \approx 30$.

More recently V. Vyssotsky [23] has found much larger numbers with high values of $\gamma(n)$. He found that $n =$12,769,884,180,266,527 has $\sigma_\infty(n) = 1271$ and $\gamma(n) = 34.2716$, and that

$$n = 37{,}664{,}971{,}860{,}959{,}140{,}595{,}765{,}286{,}740{,}059$$

has $\sigma_\infty(n) = 2565$ and $\gamma(n) = 35.2789$. He also found a number around 10^{110} with $\gamma(n)$ exceeding 36.40.

k	n	$\sigma_\infty(n)$	$\gamma(n)$	$r(n)$
1	27	70	21.24	0.5857
2	703	108	16.48	0.5741
3	6,171	165	18.91	0.5818
4	52,527	214	19.68	0.5841
5	837,799	329	24.13	0.5927
6	8,400,511	429	26.91	0.5967
7	63,728,127	592	32.94	0.6030
8	127,456,254	593	31.77	0.6020
9	4,890,328,815	706	31.64	0.6020
10	13,371,194,527	755	32.38	0.6026
Random walks model			41.68	0.6091

Table 1. Maximal value of $\gamma(n)$ in intervals $10^k < n \leq 10^{k+1}$, $1 \leq k \leq 10$.

In Table 2 we present empirical results comparing the maximal value of $t(n)$ in intervals $10^k \leq n < 10^{k+1}$ up to 10^{11} with the prediction of (2.4). Here $\sigma_{\text{max}}(n)$ denotes the number of steps taken until the maximum value is reached. Extensions of the data can be found in Leavens and Vermeulen [16]. The agreement with the stochastic model is quite good.

In Figure 2 we plot the graphs of these extremal trajectories on a double logarithmic scale. The large deviations extremal trajectory is indicated by

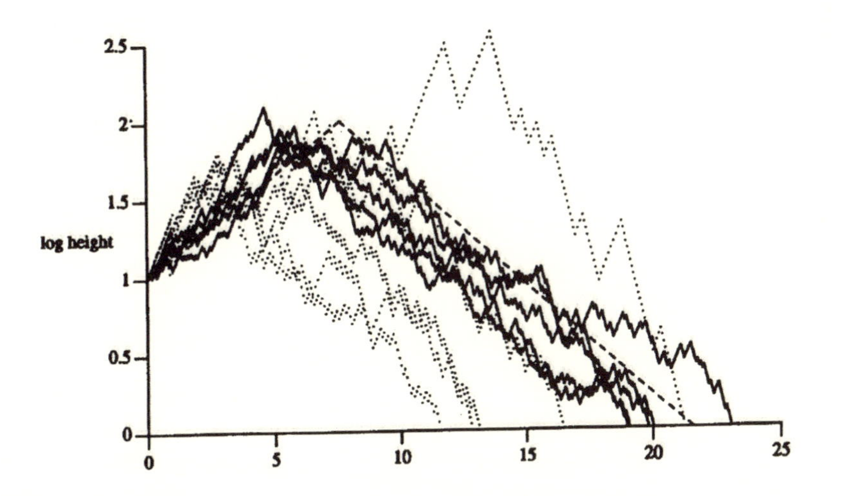

Figure 2. Scaled trajectories of n_k maximizing $(\log t(n))/\log n$ in $10^k \leq n \leq 10^{k+1}$ (dotted for $1 \leq k \leq 5$; solid for $6 \leq k \leq 10$).

k	n	$\frac{\log t(n)}{\log n}$	$\frac{\sigma_{\max}(n)}{\log n}$	$\gamma(n)$
1	27	2.560	13.65	21.24
2	703	1.791	7.32	16.48
3	9,663	1.790	2.83	12.86
4	77,671	1.819	3.46	13.14
5	704,511	1.788	2.75	11.59
6	6,631,675	1.976	5.86	23.05
7	80,049,391	1.903	5.06	19.84
8	319,804,831	2.099	4.65	19.10
9	8,528,817,511	1.909	5.20	20.03
10	77,566,362,559	1.897	6.86	19.02
Random walks model		2.000	7.645	21.55

Table 2. Largest value of $\frac{\log t(n)}{\log n}$ for $10^k \le n \le 10^{k+1}$, $1 \le k \le 10$.

a dotted line. The deviations from the stochastic model are well within the range that would be expected from the size of the expected standard deviation for such models.

3. Branching Random Walk Model

Lagarias and Weiss [15] modeled backward iteration of the $3x+1$ function by a branching random walk.

In iterating the multivalued function T^{-1}, it proves convenient to restrict the domain of T to integers $n \not\equiv 0 (\bmod\ 3)$. In this case,

$$T^{-1}(n) := \begin{cases} \{2n\} & \text{if } n \equiv 1,\ 4,\ 5, \text{ or } 7 \pmod 9, \\ \{2n, \frac{2n-1}{3}\} & \text{if } n \equiv 2 \text{ or } 8 \pmod 9. \end{cases} \tag{3.1}$$

The restriction to nodes $n \not\equiv 0 (\bmod\ 3)$ is made because nodes $n \equiv 0 (\bmod\ 3)$ never branch, so do not have any significant effect on the size of the tree; see [15] for more explanation.

The set of inverse iterates associated to any n has a tree structure, where the branching at a node in the tree is determined by the node label $(\bmod\ 9)$, using (3.1). In [15] these trees were called *pruned 3x+1 trees* because all nodes corresponding to $n \equiv 0 (\bmod\ 3)$ have been removed. Figure 3 shows pruned $3x+1$ trees of depth $k = 5$ starting from the root nodes $n = 7$ and $n = 20$; these trees have the minimal and maximal number of leaves possible for depth $k = 5$, respectively. In what follows we use the term *3x+1 tree* to mean pruned $3x+1$ tree.

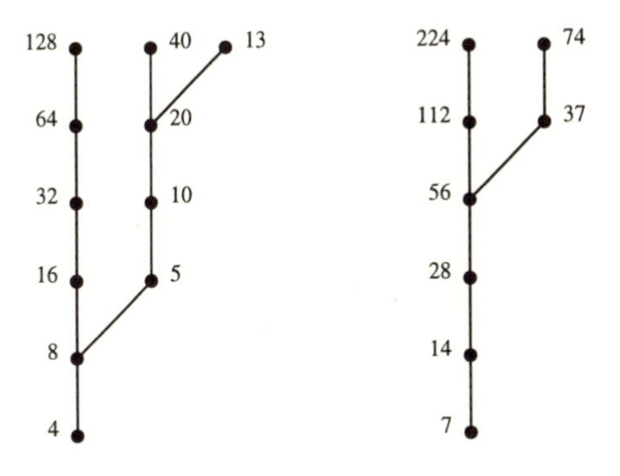

Figure 3. Pruned $3x+1$ trees of depth $k=5$ starting from root nodes $n=4$ and $n=7$.

The branching structure of a $3x+1$ tree of depth k is completely determined by its root node $n \pmod{3^{k+1}}$, hence there are at most $2 \cdot 3^k$ possible distinct $3x+1$ trees of depth k. In fact, there are duplications, and the number $R(k)$ of distinct edge-labeled $3x+1$ trees of depth k seems to grow empirically like θ^k, where $1.87 < \theta < 1.93$. (See Table 3 in Section 4.)

The problem of determining the largest trajectory to get to 1 under iteration in a $3x+1$ tree is equivalent to studying the leaf in a $3x+1$ tree having the smallest value, where the values of the inverse image nodes of a node of value n are $\{2n\}$ or $\{2n, \frac{2n}{3}\}$ depending on whether the tree has one or two branches, and the root node is assigned the value 1. We assign labels 0 or 1 to the edges, so that an edge from n to $2n$ is labeled 0 and an edge from n to $\frac{2n}{3}$ is labeled 1. The branching random walk arises from associating a walk to each path in the tree from a root node to a leaf node at depth k, where the random walk starts at the origin on the real axis at the root node and takes a step $\log 2$ along an edge labeled 0 and a step $\log \frac{2}{3}$ along an edge labeled 1. The random walk at a given leaf node ends at a position u on the real line, and the value assigned to that vertex by the rules above is exactly e^u. Lagarias and Weiss [15, Theorem 3.4] show that the expected size of $\log n^*(k)$, where $\log n^*(k)$ is the smallest value taken among all leaf nodes of depth k of a tree, satisfies (with probability one)

$$\lim_{k\to\infty} \frac{E[\log n^*(k)]}{\log k} = \gamma_{BP} \simeq 41.677647\ldots . \tag{3.2}$$

This constant $\gamma_{BP} = \beta^{-1}$, where β is the unique positive number satisfying $g^*(\beta) = 0$, and where

$$g^*(a) := \sup_{\theta \le 0} \left[a\theta - \log(2^\theta + \frac{1}{3}\left(\frac{2}{3}\right)^\theta \right]. \tag{3.3}$$

This constant is provably the same constant as (2.1), i.e., $\gamma_{BP} = \gamma_{RW}$; cf. [15, Theorem 4.1].

In order to prove the result (3.2), it was necessary to determine the distribution of the number of leaves in a tree in the branching process. In [15] it was shown that the expected number of leaves of a tree of depth k is $(\frac{4}{3})^k$, and as $k \to \infty$ the leaf probability distribution is of the form $W(\frac{4}{3})^k$, where W is a random variable satisfying

$$\text{prob}\{a < W < b\} = \int_a^b w_k(x)dx\ , \quad \text{for} \quad 0 < a < b < \infty\ , \tag{3.4}$$

and $w_k(x)$ is strictly positive on $(0, \infty)$; see [15, Theorem 3.2].

4. Distribution of 3x + 1 Trees

Applegate and Lagarias ([1] and [3]) studied properties of the ensemble of all $3x + 1$ trees. Here the *pruned 3x+1 tree* $\mathcal{T}_k(a)$ is the tree of inverse iterates of the $3x+1$ function grown backward to depth k, from a given root node a, with all nodes that have a label congruent to $0 (\text{mod } 3)$ removed.

Applegate and Lagarias [1] empirically studied the extremal distribution of the number of leaves in a $3x + 1$ tree of depth k as the root node is varied. Let $N^-(k)$ and $N^+(k)$ denote the minimal and maximal number of leaves, respectively. Since the number of leaves is expected to grow like $(\frac{4}{3})^k$, they studied the *normalized extreme values*

$$D^+(k) := \left(\frac{3}{4}\right)^k N^-(k) \quad \text{and} \quad D^-(k) := \left(\frac{3}{4}\right)^k N^+(k)\,. \tag{4.1}$$

Table 3 gives data up to $k = 30$. In this table, $R(k)$ counts the number of distinct edge-labeled $3x + 1$ trees of depth k. Based on this data, they proposed that the normalized extreme quantities remain bounded. This can be formalized as

Conjecture C$^\#$. *Let*

$$C^- := \liminf_{k\to\infty} D^-(k) \text{ and } C^+ := \limsup_{k\to\infty} D^+(k)\,. \tag{4.2}$$

Then these quantities satisfy the inequalities

$$0 < C^- < 1 < C^+ < \infty\,.$$

Applegate and Lagarias [3] compared this conjecture with predictions derived from one of the branching random walk models of [15]. The branching process is a multi-type Galton–Watson process with six types

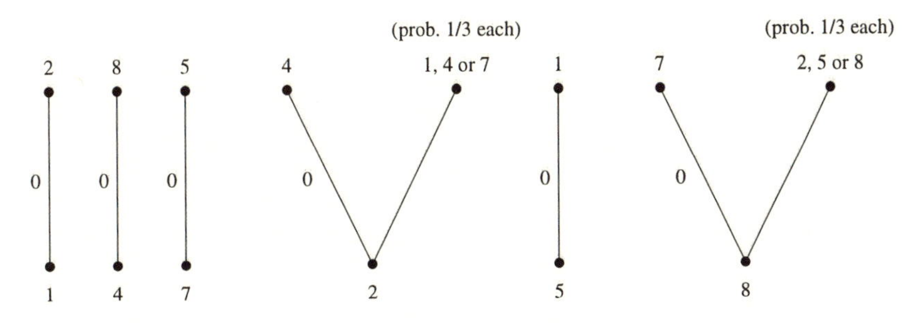

Figure 4. Transitions of the branching process $\mathfrak{B}$ [9]. The parent (bottom) always yields a child by the map $n \mapsto 2n$ (edge label 0), and it yields a second child by the multivalued map $n \mapsto \frac{1}{3}(2n-1)$ (edge label 1) if $n = 2$ or $8 \bmod 9$.

labeled 1, 2, 4, 5, 7, 8 pictured in Figure 4. (See Athreya and Ney [4] for a general discussion of such processes.)

The random walk arises from assigning the labels 0 and 1 to the edges of the resulting tree, where 0 represents the entering vertex being even and 1 represents the entering vertex being odd; the corresponding random walk takes a step $\log 2$ along an edge labeled 0 and takes a step $\log \frac{2}{3}$ along an edge labeled 1. The base vertex of the tree is located at the origin, so that each vertex of the tree corresponds to a random walk ending at some position u on the real line, and we consider the value of that vertex to be e^u. This value represents the initial value of the $3x+1$ iteration starting from that vertex of the tree. We make $\tilde{R}(k)$ independent draws of a tree of depth k generated by this process, choosing the root node uniformly (mod 9), and choosing $\tilde{R}(k) = \Theta^k$ for a fixed constant $\Theta > 1$. (For actual $3x+1$ trees, $\tilde{R}(k) \le 3^{k+1}$ so $1 < \Theta \le 3$.) We consider as random variables the smallest and largest number of leaves that occur among this set of trees. Let $\tilde{N}^-(k)$ and $\tilde{N}^+(k)$ denote the expected value of the smallest number of leaves, respectively largest number of leaves. The analogues of normalized extreme values in the models are

$$\tilde{D}^-(k) := \left(\frac{3}{4}\right)^k \tilde{N}^-(k) \quad \text{and} \quad \tilde{D}^+(k) := \left(\frac{3}{4}\right)^k \tilde{N}^+(k) \,. \tag{4.3}$$

In [3] it is proved[1] that

$$\limsup \tilde{D}^+(k) = \infty$$

and

$$\liminf_{k\to\infty} \tilde{D}^-(k) = 0 \,.$$

[1] These facts are derived using the probability density for W given in (3.4).

					$3x+1$ Function Trees		Branching Process	
k	$R(k)$	$N^-(k)$	$N^+(k)$	$\left(\frac{4}{3}\right)^k$	$D^-(k)$	$D^+(k)$	$\tilde{D}^-(k)$	$\tilde{D}^+(k)$
1	4	1	2	1.33	0.750	1.500	0.750	1.500
2	8	1	3	1.78	0.562	1.688	0.562	1.557
3	14	1	4	2.37	0.422	1.688	0.422	1.669
4	24	2	6	3.16	0.633	1.898	0.633	1.728
5	42	2	8	4.21	0.475	1.898	0.475	1.792
6	76	3	10	5.62	0.534	1.780	0.534	1.778
7	138	4	14	7.49	0.534	1.869	0.409	1.911
8	254	5	18	9.99	0.501	1.802	0.401	1.923
9	470	6	24	13.32	0.451	1.802	0.375	1.986
10	876	9	32	17.76	0.507	1.802	0.394	2.026
11	1638	11	42	23.68	0.465	1.774	0.352	2.054
12	3070	16	55	31.57	0.507	1.742	0.342	2.076
13	5766	20	74	42.09	0.475	1.758	0.307	2.118
14	10850	27	100	56.12	0.481	1.782	0.305	2.131
15	20436	36	134	74.83	0.481	1.791	0.300	2.166
16	38550	48	178	99.77	0.481	1.784	0.302	2.190
17	72806	64	237	133.03	0.481	1.782	0.289	2.211
18	137670	87	311	177.38	0.490	1.753	0.281	2.232
19	260612	114	413	236.50	0.482	1.746	0.273	2.255
20	493824	154	548	315.34	0.488	1.738	0.271	2.270
21	936690	206	736	420.45	0.490	1.751	0.268	2.292
22	1778360	274	988	560.60	0.489	1.762	0.265	2.310
23	3379372	363	1314	747.47	0.486	1.758	0.259	2.327
24	6427190	484	1744	996.62	0.486	1.750	0.255	2.344
25	12232928	649	2309	1328.83	0.488	1.738	0.252	2.360
26	23300652	868	3084	1771.77	0.490	1.741	0.249	2.375
27	44414366	1159	4130	2362.36	0.491	1.748	0.246	2.390
28	84713872	1549	5500	3149.81	0.492	1.746	0.243	2.405
29	161686324	2052	7336	4199.75	0.489	1.747	0.240	2.419
30	308780220	2747	9788	5599.67	0.491	1.748	0.238	2.433

Table 3. Normalized extreme values for $3x+1$ trees and for the branching process.

These results for the branching random walk model do not agree with Conjecture C# above. The number of leaves in actual $3x+1$ trees empirically exhibits less variability than is predicted by this branching random walk model. The paper [3] presents more evidence in favor of Conjecture

C# and formulates additional conjectures concerning the average number of leaves in $3x+1$ trees that have a fixed node $n(\bmod 3^{k+1})$.

References

[1] D. Applegate and J. C. Lagarias, Density Bounds for the $3x+1$ Problem I. Tree-Search Method, *Math. Comp.* **64** (1995), pp. 411–426.

[2] D. Applegate and J. C. Lagarias, Density Bounds for the $3x+1$ Problem II. Krasikov Inequalities, *Math. Comp.* **64** (1995), pp. 427–438.

[3] D. Applegate and J. C. Lagarias, On the Distribution of $3x+1$ Trees, *Experimental Math.*, **4**, (1995), pp. 101–117.

[4] K. B. Athreya and P. E. Ney, *Branching Processes*, Springer-Verlag, New York, 1972.

[5] T. Cloney, C. E. Goles, and G. Y. Vichniac, The $3x+1$ Problem: a Quasi-Cellular Automaton, *Complex Systems* **1** (1987), pp. 349–360.

[6] L. Collatz, On the origin of the $(3n+1)$ Problem (in Chinese), *J. of QuFu Normal University*, Natural Science Edition, **12**, No. 3, (1976), pp. 9–11.

[7] J. H. Conway, Unpredictable Iterations, Proc. 1972 Number Theory Conference, Univ. of Colorado, Boulder, Colorado, 1972, pp. 49–52.

[8] C. J. Everett, Iteration of the Number Theoretic Function $f(2n)=n$, $f(2n+1)=3n+2$, *Advances in Math.* **25** (1977), pp. 42–45.

[9] M. R. Feix, A. Muriel, D. Merlini, and R. Tartani, The $(3x+1)/2$ Problem: A Statistical Approach, Proc. 3rd Intl. Conf. on Stochastic Processes, Physics and Geometry-Locarno, 1990.

[10] Martin Gardner, Mathematical Games, *Scientific American* **226** (1972), (June), pp. 114–118.

[11] L. E. Garner, On Heights in the Collatz $3n+1$ Problem, *Discrete Math.* **55** (1985), pp. 57–64.

[12] I. Korec, The $3x+1$ Problem, Generalized Parcal Triangles, and Cellular Automata, *Math. Slovaca,* **44** (1994), pp. 85–89.

[13] J. C. Lagarias, The $3x+1$ Problem and Its Generalizations, *Amer. Math. Monthly* **82** (1985), pp. 3–23.

[14] J. C. Lagarias, Pseudorandom Number Generators in Cryptography and Number Theory, in: *Cryptology and Computational Number Theory,* C. Pomerance, ed., Proc. Symp. Appl. Math., Vol. 42, AMS, Providence, R.I. 1990, pp. 115–143.

[15] J. C. Lagarias and A. Weiss, The $3x+1$ Problem: Two Stochastic Models, *Annals Appl. Prob.* **2** (1992), pp. 229–261.

[16] G. Leavens and M. Vermeulen, $3x+1$ Search Programs, *Computers Math. Appl.* **24**, No. 11 (1992), pp. 79–99.

[17] M. Luby, *Pseudorandomness and Cryptographic Applications,* Princeton University Press: Princeton, NJ, 1996.

[18] H. Muller, Das $3n+1$ Problem, *Mitteilungen der Math. Ges. Hamburg* **12** (1991), pp. 231–251.

[19] T. Oliveira e Silva, Maximum Excursion and Stopping Time Record Holders for the $3x+1$ Problem: Computational Results, *Math. Comp.*, **68**, (1999), pp. 371–384.

[20] D. Rawsthorne, Imitation of an Iteration, *Math. Mag.* **58** (1985), pp. 172–176.
[21] R. Terras, A Stopping Time Problem on the Positive Integers, *Acta Arithmetica* **30** (1976), pp. 241–252.
[22] B. Thwaites, My Conjecture, *Bull. Inst. Math. Appl.* **21** (1985), pp. 35–41.
[23] V. Vyssotsky, private communication.
[24] S. Wagon, The Collatz Problem, *Math. Intelligencer* **7** (1985), pp. 72–76.
[25] G. J. Wirsching, The Dynamical System Generated by the $3n + 1$ Function, Lecture Notes in Math. No. 1681, Springer-Verlag, New York, 1998.